D0073458

HISTORY OF LOGIC

HISTORY OF LOGIC
IV

Österreichische Akademie der Wissenschaften, Wien

Institut für Wissenschaftstheorie, Salzburg

Gödel-Symposium in Salzburg

10-12 July 1983

BIBLIOPOLIS

Kurt Gödel
(1906-1978)

R. GÖDEL - O. TAUSSKY-TODD - S. C. KLEENE - G. KREISEL

Gödel Remembered

Salzburg 10-12 July 1983

Edited by

PAUL WEINGARTNER

and

LEOPOLD SCHMETTERER

BIBLIOPOLIS

ISBN 88-7088-141-5

Napoli, via Arangio Ruiz 83

Printed in Italy by «Grafitalia» Cercola - Napoli

Photoset by «Jet Photocomp Center» - Napoli

CONTENTS

The Gödel Family: mother, Kurt, father, Rudolf

FOREWORD

The present volume contains most of the papers given at the 1983 Gödel-Symposium in Salzburg, held at the Institut für Wissenschaftstheorie of the International Research Center Salzburg under the auspices of the Austrian Academy of Science and the Institut für Wissenschaftstheorie. It was sponsored by the same two institutions.

The idea was to bring together scholars who have known Kurt Gödel personally, and who have discussed with him scientific, philosophical and personal matters. Fortunately, it was possible to get such people who had personal contact with Gödel at different periods of his life, thereby complementing each other's accounts. His brother Rudolf who remembers their childhood, and especially their mother. Taussky-Todd knew Gödel in his student years (from 1925 on), afterwards as a doctor and «Privatdozent», mainly in Vienna. Kleene had contacts with Gödel, attended Gödel's lectures at Princeton 1933/34, of which he prepared notes together with J. B. Rosser, and kept in touch. Kreisel had discussions with Gödel after the second world war. The contributers deal with topics concerning Gödel's life and interests from their personal points of view. In particular, they report on Gödel's teachers and friends at the University of Vienna, Gödel's reception at Princeton, Gödel's views on Church's, Kleene's and Rosser's results and the development, over the years of Gödel's interests in intuitionistic logic, but also, more generally, in the foundations of logic and mathematics, and its philosophical background.

A letter by Zermelo (to Reinhold Baer) is here published for the first time. It was not known to the participants of the symposium, and shows that Zermelo remained unconvinced of Gödel's achievements after they had met and talked at Bad Elster (a matter taken up in two of the contributions).

Finally, the editors want to thank the authors for their interesting and careful analyses, in one case much more detailed than the lecture, Ulrich Felgner placing Zermelo's letter at our disposal, and for giving us permission to publish it, and the publisher, Mr. del Franco, for the excellent cooperation.

PAUL WEINGARTNER
LEOPOLD SCHMETTERER

Rudolf Gödel

HISTORY OF THE GÖDEL FAMILY

Translated by
SUSAN SIMONS

My mother's father[1], Gustav Handschuh, came from the Rhine area. His father had settled in Brünn (Brno) as a hand-loom weaver, but soon lost his livelihood as the hand-loom was replaced by machines, so the family grew up in straitened circumstances. My mother often recounted how her father had had to work hard in his youth to work his way slowly up to a good position in the then rich and respectable firm of Schöller's, where he finally had the sole power of procura, a sign of his ability and reliability. (The Schöller family were incidentally also from Dühren am Rhein and quickly made a name for themselves in the textile and sugar industries.) Our grandfather had had to acquire his education at the same time as working and often did not even have enough money to buy the necessary books, so that he was forced to copy them by hand. This hard lot was probably the reason for his developing a somewhat hard, inflexible character which was not always easy for his wife and family to bear. He provided his children, two boys and three girls, with a good education. So in her youth my mother went for a time to a French finishing school, which was in fact usual only among very well-off families. Gustav Handschuh was also very active in public life, and was among other things a founder member of the Brünn Ambulance Society (Brünner Rettungsgesellschaft) and the Institution for Retired Officials (Pensionsanstalt für Beamte), a very progressive attitude for that time.

My mother's mother came from Iglau (Jihlava), a small town in north-western Moravia, then a German-speaking enclave, as were many other places in Bohemia and Moravia, which was then gradually becoming more Czech. Judging from her stories, my mother had never felt at home in

[1] It is perhaps not unimportant to mention that my brother and I always addressed our mother as «Mama», but were nevertheless on very amicable terms with her. (Translator's note: presumably «Mama» was stressed, in the formal way, on the second syllable.)

GÖDEL REMEMBERED
Salzburg 10-12 July 1983
Ed. by P. Weingartner and L. Schmetterer
BIBLIOPOLIS 1987

Brünn and suffered from homesickness up until her death. She was a proper little housewife, who gave herself up completely to domestic life.

My mother had two brothers and two sisters. The older of the brothers was an officer but a somewhat rash character who cost his parents much money and finally gave up military service and became a civil servant. He died relatively young, a little over 50 years old.

The other brother went into the textile industry, as did many young people in those days in Brünn (the «Austrian Manchester»).

My mother's younger sister died of diphtheria at the age of six, a blow from which her mother never quite recovered. She was a somewhat strange, bright, perhaps precocious child and my mother could relate a number of her amusing remarks which were nevertheless strange for a child.

My mother's elder sister never married and spent her last years (during the Second World War) together with my mother in our villa in Brünn.

My mother's youth, spent in the peaceful times Europe enjoyed before the turn of the century, was in general happy and untroubled. She was born on 31st August 1879, was a bright, lively and jolly girl, had many girlfriends and was even then interested in sport, a very good gymnast and ice skater. She greatly enjoyed this sport in particular and even in old age she would be thrilled by the performances of figure skaters. In her youth of course such acrobatic achievements could not even have been dreamt of.

The house in which my mother grew up was built rather in the Biedermeier style with open galleries on which the neighbours would meet in the evenings for a pleasant chat, often to the sound of the Tattoo (or 'Taps': bugle solo which was played in the nearby barracks on the Spielberg at 9 p.m. when the soldiers had to return to barracks). She often spoke of these pleasant hours in later years and with her gregarious nature missed being with other people. Up until old age she kept in contact, personally or by letter, with several of the girlfriends of her youth, until the end of the Second World War, when the expulsion of the Germans from Czechoslovakia tore great holes in this circle, as several acquaintances died on the «Death March» from Brünn to Vienna. Other Jewish friends had already been taken away after the occupation of Czechoslovakia by the National Socialists; so my mother was lonelier in old age than is usually the case, which obviously caused her suffering. The circle of friends of her youth was a remnant of pleasant Biedermeier days where the families played music and put on amateur theatricals together. People knew how to enjoy celebrations, so that often several families were gathered into a large group, and it was also here that my mother met her future husband.

I ought to mention that in the years before the turn of the century Brünn was a sensitive spot in the issue of nationality. The Germans of course attempted (in the end unsuccessfully) to keep down the advancing Czech element. The intelligentsia and «high society» were then still German, so my mother was quite definitely but not exaggeratedly brought up as a German nationalist. That she was nevertheless later from the beginning a staunch opponent of National Socialism speaks for her intelligence and above all for her conception of right and humanity. (The atrocities of National Socialism became known to us, as to many others, only much later, which perhaps partly explains why many thoroughly «decent people» including acquaintances of ours went along with the movement.) In my mother's youth there were in any case frequent brawls between Germans and Czechs in Brünn and in old age she could still remember the clatter of horses' hooves in front of her house as the Dragoons were called out to separate the opposed parties.

My mother was Protestant and in her parents' house there prevailed an enlightened piety. The children always had to go to church Sundays where at that time there were preachers who were well known far beyond the boundaries of Moravia (especially Trautenberger). As her husband was attached to no religion (he was nominally Old Catholic) my brother and I were brought up as freethinkers and later too I had no proper relation to religion (unfortunately!?). How necessary the support is that is thereby taken away from one is something I first felt later when deaths and blows of fate hit our family.

I believe my mother's marriage was not a «love match» (she met her husband Rudolf Gödel because he lived in the same house). But it was certainly built on affection and sympathy. My mother was no doubt impressed by the energetic capable nature of my father,who had worked his way up from being a minor industrial employee to Director and later partner of one of the most famous textile factories in Brünn. And he, being of a more serious and ponderous disposition, must have found pleasure in her cheerful friendly nature. This nature changed in the course of my mother's life. She could still laugh heartily in old age but from middle age onwards there was a certain melancholy alongside her apparent cheerfulness, for which no cause could be found in her outward circumstances, since life improved steadily from the moment of her marriage until the early death of our father. Shortly before the First World War my father had bought a particularly well-sited garden in Brünn on the southern slope of the Spielberg (an old fortress surrounded by a large park) and here he built a villa which was finished shortly before the beginning of the war. Of course the years of the First World War were full of cares for my parents, but my father did not have to go into the Army and the events of war did not reach as far as Brünn, so our family came through in one piece. All the same my father had lost the best part of his wealth in War Loans which was a heavy blow for someone who has earned every penny himself. In the following years however the newly-founded Republic of Czechoslovakia recovered relatively quickly with the help of the Allies, so that my father was once again able to attain a high standard of living.

When we moved into the new villa in 1913 my brother was seven years old and I was eleven. We had two dogs, a Doberman and a small ratter and the latter especially was a very droll animal with which we had lots of fun and whose death had the whole family in tears.

The villa was surrounded by a fine garden in which there were many fruit trees, so in Spring it was resplendent in superb blossom. My mother was a great lover of nature and right up until an advanced age she would often speak of the first days of Spring in this garden, of the birdsong which woke her early, of the joy of finding the first violet and of listening to the evening song of the song thrushes. She never quite got over the loss of this garden in the Second World War. The reawakening of nature in Spring was always a miracle for her and even in her youth she once said «it would not be so terrible having to die, but not being able to experience Spring any more...»

Until the death of my grandfather we celebrated Christmas at home with many relations, often up to eighteen people. Beforehand we would go with our parents to our grandparents where there would be a small distribution of presents. There the Christmas tree was (according to Third Reich custom?) fixed to the ceiling of the room at the top and so could turn freely on its axis. Then at home there would be a grand distribution of presents, for my parents were very generous in this way and weeks before we were allowed to choose some of the things we wanted from a catalogue from a Vienna toyshop (which my mother did not really, like as it was «unpoetic») but there was always room for surprises.

Our grandmother prophesied a great future for my brother even as a child with whom she often played.

During the First World War a small circle of cultivated people met in our house and still occasionally played music. My mother played the piano very well and was especially interested in the art of Lieder and was a good accompanist. Even later when no more «singers» came to us she often played Lieder — both melody and accompaniment — at the piano and when eventually there were good gramophone records she was able to enjoy in particular the Lieder of Schubert, Hugo Wolf, Löwe and Richard Strauss. She always regretted that neither of her children was more interested in music and that her husband too really went to concerts more for her sake. She never missed a Liederabend with [Leo] Slezak (one of the best Lieder singers of the time). Slezak was born in Brünn and in his youth grew up in the same house as my mother where he and his mother lived in very poor circumstances until Robinson discovered his voice and had him trained.

I should here mention that my mother was a very gifted hostess who loved to have people around her and gave many a fine meal. She was a model housewife who looked after everything even when we had several servants, kept everything in order and also enjoyed cooking herself now and again. Later when I was living alone with her in Vienna she baked us every week a fine «Striezl» [a plaited milkbread] for breakfast, until the illness in her 86th year made this work impossible for her. In my whole life I have hardly known another woman who was on the one hand so culturally lively and interested and on the other hand had so much interest in and knowledge of domestic economy. Her library, which she lost in the Second World War in Brünn, was of a high quality. For many years she interested herself in the life of Goethe and owned a large number of works about the Goethe circle in Weimar. Her favourite reading included the letters of Billroth (a great 19th century Viennese surgeon) and the letters of Bismarck to his bride-to-be; in Belles Lettres she liked *Effi Briest* by Fontane which she often read in old age. As a young woman around 1900 she had much sympathy with the then modern literature and valued especially Schnitzler and P. Altenberg. She happily read fairy tales again and again, a sign of her romantic disposition. Even in old age she could recite poems by Goethe, Heine, Lenau and others by heart. As a young woman she composed poems herself which she later unfortunately burnt. As far as I can remember those few that she later recited as signs of her «youthful sins», many could have been published. I can hardly remember my mother reading serialised stories in the daily papers or crime thrillers.

After the First World War we read political and historical novels together; my mother was very interested in the lives of Franz Josef and Empress Elisabeth (without being a monarchist, but she had after all lived for almost forty years in the old Austro-Hungarian monarchy). It was a sign of her independent judgement that she always pitied the Emperor more than the Empress — who is mainly portrayed in the literature as the more pitiable — she could never forgive her for leaving her husband alone so much and behaving in what was then an eccentric way, for which the lonely monarch always forgave her. My mother formed her own opinion about everything and would defend it with a certain stubborness though she was still ready to be corrected. Her views were generally dictated by sound common sense, uninfluenced by the general opinion.

The relationship of my brother and me to our mother was always a very hearty, almost comradely one. Although my brother was perhaps less closely bound into the family and went his own way earlier, he was, especially in later years, when he was rather sickly, of particular concern to her.

We were both very good at school and gave our parents no cause for concern on this account. Even in High School my brother was somewhat more one-sided than me and to the astonishment of his teachers and fellowpupils had mastered University Mathematics by his final Gymnasium years.

My mother could tell many stories about my brother as a child which in her view even then suggested that he would become a world-famous scholar. For example when about four years old he was nicknamed «Mr. Why» by our acquaintances, because he always wanted to get to the bottom of things with his intensive questioning. She often recounted a situation which was embarrassing for the parents, when my brother (again about four or five years old), during a social gathering, tentatively examined an elderly lady for a long time and than naively asked «Why have you got such a long nose?». Sadly, many of the details that my mother could recount escape me and she herself could not be brought to write an autobiography, which would have seemed to her like a farewell to life; her whole life long she was very afraid of death, especially that she would die all alone, which thankfully did not happen.

The family's life changed decisively when my brother and I, after completing our schooling in Brünn (I in 1919, my brother in 1923), took up our studies at Vienna University. I enrolled in Medicine, my brother in Philosophy. I in particular often visited Brünn in my first student years (it normally took three hours by train to get there, though in the immediate post-war period 1919-1920 sometimes 12-15 hours!). But my mother especially found her home now somehow empty, in particular as there was no prospect of our making a career in Brünn. The prospects of making a career as a German in Czechoslovakia were not favourable after the sad end of the Austro-Hungarian monarchy, so we both had to get used to the idea of remaining in Austria. In our student years our parents often visited us and my father was also often in Vienna on business. The first years after the First World War were very sad ones for Vienna and there was not only a shortage of food but also of coal and many other necessary consumer goods. So our parents often brought us things, above all food, from Czechoslovakia which was then much richer. Somewhat later in the twenties, when things in Vienna had improved again, social and artistic life was very lively. I can remember many interesting performances in the Burgtheater and the Opera, which we visited with our parents, and also the first performances in the Theater in der Josefstadt (a favourite theatre of my mother) which had then just been opened by Reinhardt. We also got to know and love the surroundings of Vienna. We were suddenly torn out of this pleasant and varied life by a severe illness of my father, who was then only 54 years old, and this illness led to his death in February 1929. Our whole world collapsed especially for my mother. For months my mother in particular was in a condition which caused us to fear the worst. We could not possibly leave her alone in Brünn in this condition, so the villa and garden were let and my mother, my brother and I moved into a large apartment in the Josefstadt, the Doctors' District of Vienna. There was also an old lady, «Aunt Anna», living with us who had lived in the villa in Brünn with my parents for many years. She was my father's foster mother. He was born in Vienna, but grew up in Brünn; his mother had entrusted him to relatives after the death of his father (who committed suicide at an early age?). Out of gratitude, my father had taken his «aunt» into his house when she was later reduced to poverty. She was an altogether good natured old lady, it was only that she sometimes got on my mother's nerves with her pessimism and excessive anxiety about everything. But in general the relationship with this «mother-in-law» was a good one, though as usually happens in such cases, it perhaps did not always have a good influence on my parent's marriage in later years.

After my mother's condition had slowly improved there followed a number of years in Vienna which will always hold pleasant memories for us all. My mother slowly entered more fully into life and as Vienna in the thirties had many fine theatre performances and concerts to offer we and my brother often enjoyed pleasant evenings together, which concluded with long discussions about what we had seen. At that time my brother was very interested in the theatre; sadly, it was later obviously different.

I had completed my studies before the death of my father; my brother now also had his Doctorate and was made a Docent at the University of Vienna at a very early age on the strength of a scientific work. At that time I was working at the famous Vienna Wenckebach and Eiselberg Clinics and finally settled on Radiology at the Holzknecht Institute.

My brother began at that time to be well-known in scientific circles and we frequently had famous scholars from home and abroad as visitors. Nevertheless we did not then grasp the magnitude of his discovery and had to have it pointed out to us by others, for my brother hid his light under a bushel and did not like speaking about it.

The year 1933, in which the National Socialists came to power in Germany, holds no particular memories for me in Vienna, as none of us was very interested in politics, so we could hardly judge correctly the significance of this event. Two events quickly opened our eyes: the murder of Chancellor Dollfuss and the murder (by a National Socialist student) of the philosopher Professor Schlick in whose circle my brother had moved. This event was surely the reason why my brother went through a severe nervous crisis for some time, which of course was of great concern, above all for my mother. Soon after his recovery he received the first call to a Guest Professorship in the U.S.A. This event and the fact that life in Vienna was becoming even more expensive for us, whose wealth was in Czechoslovakia, was the reason why my mother returned to her villa in Brünn. That was in 1937, one year before Hitler occupied Austria.

Thus it was that my mother spent the period of the Second World War in Brünn. That had the advantage that she was better supplied with food and could often help me out but it was of course a heavy burden for her to be entirely dependent on herself particularly in those uncertain and disturbed times. Luckily as Director of a large Radiology Institute I did not have to go into the Army so my mother was often able to visit me; I greatly admired her brave perseverance in this difficult time. Living among the Czechs at that time meant being surrounded by enemies who were only waiting for the defeat of Germany at the end of the war in order to unleash their pent-up hatred. «Aunt Anna» had died at the beginning of the war (in an agitated struggle with death that lasted days) and also my mother's sister, who had lived with her in the villa during her final years, died in 1942.

Although most of my mother's acquaintances had become National Socialists, from the first moment my mother was a vehement opponent, and caused me a lot of concern with her frequent careless remarks. Luckily she was not reported by anyone and so escaped concentration camp. There were difficulties with the villa's Czech housekeeper, because the Czechs' hatred of the Germans was at that time of course great, as there were frequent executions of Czech «traitors», notice of which was posted up on red notices in the streets.

I was able to visit my mother every now and again during the war and in August 1944, when the situation in the theatres of war had become very serious for Germany, had the happy idea of bringing her to Vienna and not allowing her to return to Brünn. She surely owed her life to this circumstance as with the expulsion of the Germans from Brünn most of her contemporaries lost their lives and only a few reached Vienna, where they were received with mixed feelings, in a semidevastated and starving city. Many of the victims amongst the refugees died of typhoid which was then rife among them; and it is a miracle that there was no epidemic at that time in Vienna. All our relatives who were still living in Czechoslovakia were then driven out in some cases after going through a terrible time in concentration camps. Most went to Germany and those who were not too old to do so started a new life there. There was no distinction made in Czechoslovakia between those who were members and sympathizers of the National Socialist Party and those who were not and to judge from the stories of eyewitnesses it was reminiscent of the worst horrors of the Thirty Years' War, the expulsion of the Huguenots, or the expulsion of the Protestants from Austria at the time of the Counter-Reformation; indeed I believe these horrors were even surpassed.

So my mother lived through the end of the war with me in Vienna and kept going bravely despite our diffuculties and also the air raids at the end. My brother had married in 1938 (his wife Adele Porkert was Viennese) and was briefly in Vienna at the beginning of the war, but he was fortunately able to return via Russia and Japan to the U.S.A. and spent the war there.

My mother and I spent the last three days in the cellar while the battle for Vienna raged and then the Russians were there and the war was over for us. We came back from our first walk in the inner city of Vienna in tears; we found everything so devastated; the roof of St. Stephen's Cathedral collapsed, the Opera and Burgtheater burnt out, ruins and rubble everywhere, piles of shells on the streets, and corpses which had still not been buried. Even though the destruction of Vienna was not as extensive as in many German cities, we still had the impression that it would take decades to rebuild everything; thank goodness this was not the case and the rebuilding went much faster than we had expected.

Luckily my apartment and surgery had remained intact and I was soon able to begin my work again. Shortly afterwards the first telegram came from my brother and then also food parcels arrived from him which helped us through the worst time after the war. We were all so happy that the war was over that we willingly put up with the early difficulties of the post-war period. We later remembered laughingly how my mother had once cooked apricot dumplings in my surgery on a small cooker (which was quite a technical feat), as the electricity was not functioning in my apartment but was working in my surgery. After the heavy burden of the war was lifted from us these things were taken in good humour.

So from August 1944 my mother lived with me in my small bachelor apartment; our property in Czechoslovakia was lost. She was then 65 years old and still hale and hearty, and until her accident when she was 85 years old remained mentally and physically in exceptionally good shape, so that during these years she was able to make numerous car journeys with me sometimes as far away as Switzerland and Italy, in addition to several visits to her brother (who had also fled to Germany) and four flights with me to the U.S.A. to visit her son. So in her youth she had travelled by horse-drawn tram and in old age flown by jet to the U.S.A.

In particular these trips to the U.S.A. were always an occasion for her. It was not just that she was understandably proud of her famous son, but also that she felt particularly happy in his household, in his house and garden being looked after by him and her daughter-in-law. She also openly and without prejudice welcomed everything that was new and often better in the U.S.A. She admired the magnificent facilities of the University and the Institute of Advanced Studies in Princeton where my brother was Professor. She greatly enjoyed going shopping in the Shopping Center, an institution that at the time of our first trip to America we still did not have at home. She was also very impressed by the World Exhibition that we visited in 1964 and I think if there had been a possibility of moving with me to the U.S.A. she would have gone with pleasure despite her age.

My mother had remained so young at heart and so open to new ideas that no-one saw her as being as old as she was and we too never thought that this would have to change sometime. At 85 she was still more quick-witted than many younger acquaintances and she had retained her love for nature, so every journey gave her unadulterated pleasure especially those to the Salzkammergut which she greatly loved. If it had not been for an increasing hardness of hearing (which could unfortunately not be helped by a hearing-aid), she would certainly have kept going to the theatre or to concerts in her last years, something which was now sadly denied her. In the last years my mother and I were very much on our own, which did not worry me much as I was kept very busy with my work. My mother found it much harder and obviously also felt the lack of grandchildren and that the family would die out with her two sons.

Shortly after our fourth trip to America in July 1964 my mother had such a bad fall in the apartment that she broke her right arm at the shoulder. This caused not only severe shock but also a long period of pain and sleeplessness (and the need for painkilling drugs) which sapped her reserves of strength and was probably the indirect cause of a heart attack which she had six months later. This too she got over remarkably well. After six weeks in bed she was sufficiently recovered to be allowed to walk again. Later she could go out again and once again undertake car trips in Austria. However at intervals she had repeated attacks of angina pectoris, of varying severity. She was very sad to be unable to fly to the U.S.A. in April 1966 for the sixtieth birthday of her son. (As a small recompense we spoke to my brother by telephone on his birthday.) She bore her illness with great patience but often complained of having to suffer from such an insidious ailment, with unpredictable attacks which could not be prevented. In June 1966 we were still able to spend a very pleasant time in Ischl in the Salzkammergut together with a good friend. Our last trip was to St. Wolfgang where on the famous terrace of the White Horse Inn we sat in glorious weather until the last rays of the setting sun.

Soon after our return to Vienna her condition became markedly worse, the heart attacks came every day and on 23rd July 1966 our dear mother succumbed to a severe attack and left us forever. We found this all the harder to grasp as she had been right to end mentally fully alert and interested in everything. Her best friend and I (as well as her physician) were at her death bed so her fear that she would have to die alone happily did not come true. A morphium injection allowed her to pass away in peace.

My Father's Family (Supplement: March 1978)

I know only a few facts about my father's family.

Our father's father came from Brünn but later lived with his wife,who was Viennese, in Vienna, and was employed in the leather industry. I never met my father's parents; in our family we hardly even spoke of them. My father's father apparently committed suicide at the end of the last century.

My father (born 1874) was brought up not by his parents but by a sister of his father in Brünn: «Aunt Anna» who later lived in our parents' villa and only died during the Second World War.

In Brünn my father began studying at Grammar School but showed no talent and no interest in a «classical education», so he was sent to a weaver's school at about twelve.

The «clothmaking» trade, whose centre in Austria-Hungary was in fact Brünn, interested him so much that he completed his study at the school with distinction and immediately obtained a position in the then very renowned textile factory of Friedrich Redlich. He worked in this factory until his death as Director and later as partner. He contributed much to the excellent reputation of this firm.

My father was on bad terms with his mother because she and her relations made too many pecuniary demands of him at a time when he himself earned little and had yet to establish himself.

My father had two sisters living in Vienna. One was unmarried and had devoted herself to the sect of Jehovah's Witnesses. The other was married and my parents and I were sometimes invited round during my student days. They played music: their son had a very nice voice and sang Viennese songs and folk songs accompanying himself on the lute. Another of my father's relations was Carl Gödel, an Academy painter. We — that is my Mama and I — were often invited round during the war. His wife was a teacher and they were both very well-educated and well-read. He was a giant of a man (despite surviving tuberculosis in his youth), had according to Mama a very fine voice in his youth and later played and sang a good deal at home with his wife and with acquaintances. He and his wife died during the Second World War.

Our father was a man of thoroughly practical disposition who bestowed a superior education on us children. Personal contact was perhaps a little less warm than with our mother, but he was a good father who fulfilled many wishes of us children and later provided plentifully for our studies in Vienna.

He died in 1929 at the age of 54 of an abcess of the prostate.

Supplement Feb. 1978

As far back as I remember I always got on well with my brother. The games we played together in our parents' house were almost always free of conflict. They were mainly quiet games: bricks, model trains, also board games and chess now and again. But as a small boy he was a bad loser, who could cry intensely when he lost.

It was only in the last year of his life that my brother wrote me no more letters as was also the case with our relatives or acquaintances with whom he occasionally exchanged letters. According to his doctor he was by that time already depressive and at times troubled by feelings of inferiority. At the time he already had problems with bladder and kidneys, which was perhaps partly to blame for these mental complaints.

At school my brother was always an excellent pupil who found learning easy but took very different degrees of interest in the various subjets. Mathematics and languages ranked well above literature and history. At the time it was rumoured that in the whole of his time at High School not only was his work in Latin always given the top marks but that he had made not a single grammatical error.

In these years at High School it was clear that my brother was not so attached to the family as I was. Whereas I almost always joined in the Sunday walks or outing my brother often did not, preferring to stay at home with a book and this often annoyed my father. He could appreciate nature but did not show the real love of nature that my mother possessed to such a high degree. Later when he owned a house with a beautiful garden in Princeton, he was more often to be found in his gloomy study than in the garden.

In the years 1929-37 we lived together with our mother (our father was already dead) in a large apartment in Vienna, very near the famous Josefstadt theatre. During these years he was often out in the evening taking part in meetings particularly with Menger, Schlick and Wittgenstein, but he still had a lively interest in the theatre and as far as I remember also in history (interests which he lost in the U.S.A.). After the excellent performances in the Josefstadt theatre (at that time, under Max Reinhardt, perhaps the best theatre in the German-speaking world) we always had interesting discussions with our intelligent and very well-read mother. My brother had a very individual and fixed opinion about everything and could hardly be convinced otherwise. Unfortunately he believed all his life that he was always right not only in mathematics but also in medicine, so he was a very difficult patient for his doctors. After severe bleeding from a duodenal ulcer (when I believe his life was only saved by several blood transfusions) for the rest of his life he kept to an extremely strict (over-strict?) diet which caused him slowly but surely to lose weight. He then gave up all further visits to conferences and after the war never again came to Europe.

In important matters my brother was extremely correct, very friendly and helpful by nature. During my stays in the U.S.A. he came with me several times to the Public Library in New York to help me look for literature on a particular subject. In so far as I can judge my brother's political attitude, he was naturally in favour of Democracy and Freedom.

Among his acquaintances in Princeton he was closest to Prof. Morgenstern with whose family we in Vienna were also friendly. He was also particularly affected by the death of Prof. Einstein.

As far as I know my brother had little interest for Belles Lettres, but on the other hand one could debate with him about fine arts. He had a thoroughly positive opinion of modern painting (in particular American), something which could hardly have been anticipated from his education before the First World War. In New York he and I went several times to the Metropolitan Museum, Museum of Modern Art and Guggenheim Museum. I do not know whether he had a preference for any particular artist. My brother liked listening to music, in particular light opera and Viennese operetta.

The older my brother became the more he concentrated his interest on his own subject and suppressed everything else.

My brother was of an unstable mental disposition. Even as a child of four or five he suffered for a long time from states of anxiety whenever our mother left the house. (He always got on best with his mother.) After the murder of Prof. Schlick he had a «nervous breakdown» and was in a sanatorium for some time. Later too there were occasional mental crises.

I would not presume to pass judgement on my brother's marriage.

Olga Taussky-Todd

REMEMBRANCES OF KURT GÖDEL

I thank G. Kreisel profoundly for encouragement and for sharing his wisdom with me. I also received valuable information and criticism for my Salzburg lecture from my colleague A. S. Kechris.

As a rule, a human being of achievements can only be judged by these achievements. In principle it is immaterial whether he (or she) was aided by beneficial or hampered by adverse circumstances. Famous examples show that great strength of character has helped some people overcome great difficulties. There is the German saying: 'Das Schicksal kann den Helden zwar zerschmettern, doch einen Heldenwillen beugen kann es nicht' (I do not recall where I read it). Hence there are two aspects to consider when one studies the history of a human being. However, history is a tricky, unscientific subject, depending on random events. Even a full-fledged biography is a function of its writer, however honest he may be. This write up is not a biography of Gödel in any way. It is a collection of remembrances gathered by the writer who was a colleague of Gödel at the University of Vienna (Wien), starting for her in 1925, as well as contributions from others. The main source is the most useful Royal Society obituary notice by G. Kreisel, Biographical Memoirs of Fellows of the Royal Society 26 (1980) 149-224 (to be referred to as ON).

In spite of my interest in logic, my knowledge is minute in comparison to the other contributors of this volume. Hence no details of Gödel's achievements will be included.

I myself have withdrawn from logic a long time ago. My main subject has been number theory at all times. However, I met Gödel at a logic seminar in connection with a remark of mine. I met A. Robinson after the war in England in connection with algebra and aerodynamics. But later in California I was able to stimulate him into the use of profinite ideas - he acknowledged this in a lecture and in a Bulletin article. Next, I am devoted to the study of Sums of Squares, a subject of great interest to logicians. Here I have contacts with G. Kreisel.

Kreisel in ON tells much about Gödel, but there is still much left to explain. One thing concerns money. Apparently his father died young leaving the family well provided financially. But it is not clear whether

GÖDEL REMEMBERED
Salzburg 10-12 July 1983
Ed. by P. Weingartner and L. Schmetterer
BIBLIOPOLIS 1987

Kurt was given his own share of the fortune, or the mother all of it. Kurt entered the university in 1923 and it is my guess that until he had his first invitation to Princeton he did not earn a penny, while his older brother Rudolf was a medical doctor. Then again he presumably had no income until the second invitation. He then stayed in Wien as unpaid Privatdozent until he left Austria for good. None of this is my business to know, of course. There is further the role of the mother in this family, the mother of two sons, to be considered. Was she able to give each of them the same amount of affection?

Gödel made two trips to Princeton. Several colleagues came to see him off at the station (the Westbahnhof) where he boarded a Wagon Lits compartment on the Orient Express. A fine looking gentleman, presumably his Doctor brother, stood apart from us and moved away as soon as the train started, while we others waved a little longer. Later I mentioned that morning at the station to Gödel, and he confided to me that this was not his actual departure from Vienna. He had been taken ill before reaching his boat, took his temperature and decided to return home. However, his family persuaded him to try again. I do not recall anything he told me about Princeton or the U.S.A., apart from the fact that the steaks he ate were very small — they must have been filet mignons! Both steaks and filet mignons are not really Austrian dishes.

During his second stay in Princeton his health broke down. I heard about this in Cambridge about 1936, via a student of Miss Stebbing, a Professor of Logic in London, who had heard it from Schlick and Schlick was supposed to have been pessimistic about Gödel surviving much longer. I was truly upset when this was reported.

There is no doubt about the fact that Gödel had a liking for members of the opposite sex, and he made no secret about this fact. Let me tell a little anecdote. I was working in the small seminar room outside the library in the mathematical seminar. The door opened and a very small, very young girl entered. She was good looking, with a slightly gloomy face (maybe timidity) and wore a beautiful, quite unusual summer dress. Not much later Kurt entered and she got up and the two of them left together. It seemed a clear show off on the part of Kurt.

That same girl changed quite a bit later; maybe she became a student. She came to talk to me occasionally, and she complained about Kurt being so spoiled, having to sleep long in the morning and similar items.

Apparently she was intersted in him, and wanted him to give up prima donna habits. Quite a bit later she handed a write up to Professor Menger; something on topological spaces. Early on in my life many people have handed chores over to me, chores of all sorts. This is still the case, and I still do not know what to do about it. Hence, Menger asked me to check her work. It really was not in my line. The best I could do was to sit down with the girl and read it with her, making her explain it to me. It appeared soon that she was unable to do so, but was truly grateful to have somebody to talk to about it. It also appeared that she wanted to show Kurt that she could do something.

Kurt had a friendly attitude towards people of the Jewish faith. And once he said out of the blue that it was a miracle (Wunder in German) how, without a country, they were able to survive for thousands of years, almost like a nation, merely by their faith.

Gödel was well trained in all branches of mathematics and you could talk to him about any problem and receive an excellent response. If you had a particular problem in mind he would start by writing it down in symbols. He spoke slowly and very calmly and his mind was very clear. But you could talk to him about other things too and his clear mind made this a rare pleasure. I understand that Einstein had many conversations with him.

Gödel and I were born in a part of the Austro-Hungarian empire which is now called Czechoslovakia. This part was called Moravia (Mähren); Gödel's birthplace was Brno (Brünn) and mine Olomouc (Olmütz). Gödel's family moved to Vienna in 1923, while my family had gone there already in 1909, later to Linz, then back to Vienna. Gödel entered the University of Vienna in 1923. Our teachers were three full Professors (called ordentliche Professoren): Furtwängler, Hans Hahn, Wirtinger; one ausserordentlicher Professor: Karl Menger; several Privatdozenten: Lense, Helly, Walter Mayer, Vietoris.

Only through Kreisel's ON did I learn that Furtwängler's course on class field theory almost lured Gödel into this subject. How class field theory could have profited from a man like Gödel! However, elementary number theory was an essential ingredient in Gödel's work.

Gödel did not have contact with all these teachers and scholars. But since they were the background of his mathematical education, I will describe them slightly. Furtwängler, my own teacher, had earned recognition through his early work in geodesy — I suppose during World War I.

Later he emerged as the man who proved, and disproved, Hilbert's conjectures in class field theory. He was self taught and not, as is usually assumed, Hilbert's student, in fact the two men never met. His first appointments were in Germany from where he stems. He finally had an offer from Vienna, as successor of Mertens. He had many thesis students on all levels.

Hans Hahn was Professor of Geometry in the widest sense. He was a very active person. He was politically active for the socialist party, he wrote many papers in mathematics, he was a member of the Wiener Kreis and very much interested in logic. But he was also an ardent follower of ESP. I myself attended one of his lectures on this subject where he fended off people who asked doubting questions. He was extremely adverse to fake media who harmed the relevant research. Since Gödel was not only Hahn's student and had himself apparently an inkling of related nature, there may have been conversations between the two men in this connection, but I would not know about this.

Hahn had done a great deal of work in pure mathematics. He had started off as a young man in a university position in Poland and was attached to the topology school there, but his contributions were in other branches of mathematics too. There are only three items with which I personally had active contact: his topological characterization of the Streckenbild, a subject for which he wished to be remembered, his article Über die nichtarchimedischen Grössensysteme, Sitzungsber. Wiener Akad. Math. Nat. Klasse, Abt. II a 116 (1907) 601-653; finally the work which led people later to use the term Hahn-Banach theorem in functional analysis. Clearly, Hahn had not asked for this term. I know that he would have objected to it. I was Hahn's assistant in my last year in Vienna (1933/4) which was also Hahn's last year of life. He was already very ill and I practically supervised a Ph. D. thesis for which he was the official supervisor. This thesis concerned sequence spaces and used earlier papers of his which were based on work of Helly. He made no secret of this and, in fact, he asked me to be sure to cite Helly when I gave a lecture about the results of that thesis.

I know practically nothing about his work in logic, but he was Gödel's thesis man and I expect he was highly competent on his achievements. I was present at a lecture on logic he gave to an audience of non experts. There he stressed the fact that mathematical work was nothing but a set of tautologies. This made me unhappy, for I am a devoted number theoretician and my feelings were expressed in my poem by the words «number theory is greater than what comes later in the strict athletics of mathematics», Primary Press 1979.

I know less about Wirtinger, whose subject was analysis and who was an expert on abelian functions. He was much admired by H. F. Baker in Cambridge who had invited him out there at some time. When I started my student days, the lay people in Vienna saw him as an 'Einstein'. Later when Furtwängler also proved the group theoretic part (possible through Artin's ideas) of the Hauptidealsatz and received a big prize for this, Wirtinger moved into the background. He moved towards retirement, he cancelled his lectures frequently, his hearing was almost nil and I think he became a disappointed and even angry person. I personally do not know of any link between him and Gödel.

Menger was deeply interested in Gödel's work; he moved somehow away from mathematics to logic. Initially he was a student of Hahn and had great interest and achievements in topology. He was Hahn's favourite student and the envy of Helly, Lense and W. Mayer.

I do not know much about Lense, a very kind man, who moved to München soon. Helly was a very talented man who bragged that his achievements were not numerous, but all of interest to others. During my student days W. Mayer lectured on n-dimensional differential geometry and thus became attached to Einstein who took him along to Princeton. Menger was able to do very much for Gödel, through his journal and through his private seminars held in his apartment, one evening every week.

I had no contact with Vietoris, a man who is considered the founder of algebraic topology through his early work and whose name is attached to work in algebra nowadays. I do not know about any relation between him and Gödel. He, however, constantly confused me with another girl student who did attend classes by him.

It was a seminar of Schlick that got me into contact with Gödel already in my first year at the university. In this seminar we studied Bertrand Russell's book 'Introduction to mathematical philosophy' in German translation. In the first meeting of this seminar the subject of axioms was discussed, in particular the axioms for number theory and the axioms of geometry. In the second meeting I dared to ask a question, namely about a connection between the two systems of axioms. This led, unexpectedly, to a long discussion between participants who enjoyed making lengthy speeches. Schlick never said a word, and I myself was unable to follow much of it. At the end of the meeting Schlick asked: who would like to report on this seminar next time? Nobody replied. Then Schlick asked: who will report on this seminar? Then Gödel volunteered. He started with the words: Last week somebody asked the following question...; he did not know my name, of course. However, from then on I existed for Gödel and it remained that way through my years as a student in Vienna, then after my return from Zurich, through my years as a postgraduate when I disappeared to Goettingen, later Bryn Mawr, Girton College, Cambridge, returning only during summer vacation. And when I did not run into him in the seminar it seemed quite natural to me to phone him at his residence and have a little chat. Even much later when I visited Princeton in 1948 we renewed our friendship instantaneously.

Let me return to our years as students. It became slowly obvious that he would stick with logic, that he was to be Hahn's student and not Schlick's, that he was incredibly talented. His help was much in demand. One day I told him of my new system of axioms for groups. He reproved its validity immediately. Menger asked him to study one of my papers. He saw immediately that there was only one case of the discussion that could lead to difficulties. He seemed a rather silent man, but he offered his help whenever it was needed. Austria was not a very prosperous country in those days. Hence the mathematical seminar enriched its library by publishing the journal 'Monatshefte'. They had thus a lively exchange with other mathematical journals in many countries. Further, by including reviews of books which the local reviewers had to return to the library, they acquired many books. Gödel took on a number of these reviews. In addition he was an editor of Menger's journal 'Ergebnisse eines Kolloquiums'. I myself worked on my own dissertation. Shortly before I received my doctorate, my family moved back to Vienna. I was then allowed to invite colleagues to tea (Jause). Gödel enjoyed these invitations. But he was very silent. I have the impression that he enjoyed lively people, but not to contribute himself to nonmathematical conversations.

On the whole, these tea parties were for colleagues of my own age group, or for visitors to Vienna whom I had met and whose work was not unfamiliar to me. They were either class field experts like Iyanaga from

Japan, a student of Takagi, or attached to Menger's school and Menger himself joined too. There was also Midutani, a statistician, but I do not know how he came to join Menger's group. But, as usual, I was asked to check his work. Most of these visitors came from the U.S.A. during sabbaticals or from Japan.

I had not realized that Gödel, who clearly seemed to enjoy himself, felt superior to this circle, as he in fact was. He had somehow heard that I had invited Hahn and the great Takagi, on his visit to Vienna after the Zurich Congress, to meet my family and he remarked that I counted him among the 'minores gentium'. This is a medical term for doctors of not the highest standing, who of course, could not command the highest fees. (In Austria doctors still converse in some sort of medieval Latin slang, partly to prevent patients from understanding them.)

In due course I heard that Gödel's dissertation 'Über die Vollständigkeit des Logikkalküls' was an extremely important achievement and that Schlick was very much impressed with his philosophicum — the one hour examination in philosophy which forms part of the D. Phil. examination. He achieved the position of Privatdozent (so-called Habilitation) at a very early date with a paper called 'Formal unentscheidbare Sätze'. From Auguste Dick I received the information that Hahn wrote in his recommendation... 'eine Lösung ersten Ranges, die in allen Fachkreisen grösstes Aufsehen erregte und — wie sich mit Sicherheit voraussehen lässt — ihren Platz in der Geschichte der Mathematik einnehmen wird'. This did in fact happen and shows Hahn's correct appreciation. Already in 1931 he achieved world wide fame, only one year after achieving his doctorate. He had proved the existence of undecidable mathematical statements and gave a shock to the world of logicians and mathematicians. But he had achieved even more: he had shown something that any person with a minimum of background can understand. This is a very important fact in my opinion.

Gödel had achieved many other results, perhaps not less important ones, but this is the one mainly associated with his name. The book by Hofstädter 'Gödel, Escher, Bach' became a best seller.

I saw Gödel at the 'Tagung der deutschen Mathematikervereinigung in Bad Elster' in 1931. It was there that he met Zermelo. I think that perhaps no other person is alive who remembers this event. I had good reason to know about it for I worked then with the number theoretician

Arnold Scholz, who was a great class field theory expert. Both Scholz and Zermelo worked in Freiburg. Scholz was anxious to help Zermelo and thought a discussion with Gödel would achieve this. But Zermelo was a very irascible person. He had suffered a nervous breakdown, felt ill treated, but had actually recovered at that time. He had no wish to meet Gödel. A small group suggested lunch at the top of a nearby hill, which involved a mild climb, with the idea that Zermelo should talk to Gödel. Zermelo did not want to do so and made excuses: he did not like his looks (he actually had not met him); the climb was too much for him, there would not be enough food if Gödel came along too. However, when Gödel came to join the two immediately discussed logic and Zermelo never noticed he had made the climb.

The trouble with Zermelo was that he felt he had already achieved Gödel's most admired result himself. Scholz seemed to think that this was in fact the case, but he had not announced it and perhaps would have never done so. It is not impossible that others too had some of Gödel's results. This reminds me personally of the criticism against Hilbert's work in class field theory. People nowadays feel that it was not as original as Hilbert presented it. It seems to me that it needs an energetic person to take over and with full knowledge of the work that is in the air to give creative guidance for the future. I doubt that Zermelo was such a person.

The peaceful meeting between Zermelo and Gödel in Bad Elster was not the start of a scientific friendship between two logicians. In ON Zermelo's correspondence with Gödel is (partly) available, see p. 221. Kreisel in ON, p. 169, footnote, says what Zermelo had proved, and what he had overlooked.

Gödel suffered not infrequently from severe mental breakdowns. I do not know whether they were caused by the overstrain he suffered through the creative processes he made his brain carry out, or whether they were in his set up. I felt honoured when his brother the doctor, came to the telephone when I tried to call Kurt during one of these attacks and discussed his anxiety about Kurt with me. In addition Kurt had some physical ailments.

But Gödel did not spare himself — already visible by his performance in the Schlick seminar —, he was convinced of the value of his ideas and made sure that they were known and appreciated.

Physically, Hilbert must have been the stronger one, but in his sixtieth

Philipp Furtwängler
(1869-1939)

Hans Hahn
(1879-1934)

Wilhelm Wirtinger
(1865-1945)

Karl Menger
(1902-1985)

year he had a breakdown (I think), he acquired pernicious anemia and by the time I met him (when he was 70 years old) he discussed his health practically all the time. I heard him say that mathematicians must live on a very careful diet, a fact which sounds wise.

Auguste Dick has supplied me with an amusing remark by Furtwängler concerning Gödel's result when the latter had one of his paranoia attacks: 'Is his illness a consequence of proving the non provability or is his illness necessary for such an occupation?'

Gödel gave me copies of his first two fundamental publications. Unfortunately, we lent them to a colleague from whom we could not extract them later.

In spite of my admiration for Schlick, I myself left his seminar and even his private circle to which I had been admitted. I was disappointed that these gatherings could not give me guidance for my work in number theory. Had I realized what Gödel would achieve later I would not have run away. For Gödel's results show that logic is not a subject that stands alone and is a basis for mathematical thinking, it is in fact part of mathematics. Much later A. Robinson, by reinterpreting Gödel's results and others was able to use methods of mathematical logic, see ON p. 150.

While Gödel is best known to the general public for the result on nondecidability, this achievement plays little role for the mathematician who merrily tries to prove whatever yields to his prowess. It is the use of logic as a mathematical tool that is the great novelty and is employed more and more. E.g. at the International Conference on Logic in Salzburg 1983 A. Macintyre presented his work in this area and from G. Kreisel I learn about its use in the study of the Hasse norm theorem.

Hilbert, of course, was a driving force also in logic, before Gödel took over, guided by Hilbert's unsolved problems. That logic became a part of mathematics is due to Hilbert's influence, as well as to Skolem, Herbrand (who was also the creator of the modern presentations of class field theory) and others, before Gödel.

Hilbert had a big plan. Already in 1901 his famous mathematical problems appeared, not isolated questions, but questions which penetrated the whole edifice of mathematics and became the seeds of a large growth in the world of mathematics. In particular his first two problems concern the foundations of mathematics:

1. Cantor's problem about the cardinal of the continuum.

2. The consistency of the axioms of number theory.

In 1974 Donald Martin and Georg Kreisel reported about the success concerning these problems up to that time.

In the years 1918, 22, 31... Hilbert alone or jointly with Bernays developed a large program for the foundation of mathematics. He was particularly interested in the axioms for number theory, in geometries, in the concepts of logic.

These problems then became the basis for Gödel's work although he was critical of Hilbert's claims and aims. He spoke to me about this, I think in Zurich, and lashed out against Hilbert's paper 'Tertium non datur' Goettinger Nachr. 1932, saying something like 'how can he write such a paper after what I have done?' Hilbert in fact did not only write this paper in a style irritating to Gödel, he gave lectures about it in Goettingen in 1932 and other places. It was to prove Hilbert's faith. In Hilbert's collected works, volume 3, this paper is not reproduced, but only listed as a paper not included in that volume.

Gödel, however, as said already, proved the Completeness Theorem, the non decidability; and had great success with Hilbert's first problem.

Although Gödel's home was in the mathematics seminar, although his wish was to become a student of Hahn, although he was not a member of the Wiener Kreis, he was nevertheless an offspring of the Wiener Kreis to which Hahn, Menger, Carnap, Waisman and others belonged. Remember it was Schlick's seminar where the contact between Gödel and myself commenced. The Wiener Kreis, a successor of the Mach Verein, was a creation of Schlick. Time was ripe for such a creation in the late twenties. For, the necessity to test the foundation of mathematical thinking, the methods of proofs, the axioms, the rules, became pressing. It was a time of 'Sturm und Drang'. There were Peano's axioms for the integers, as they are described in Russell's 'Introduction' and the edifice of the Principia Mathematica above them. One found a way of explaining the 'Paradoxes'.

Wittgenstein was the idol of this group. I can testify to this. An argument could be settled by citing his Tractatus.

However, one felt that not enough had been achieved. One spoke of Philosophy of Mathematics — this was also the name for Schlick's seminar. For Russell the whole of mathematics was logic. But then Gödel turned up and became the chief creator of mathematical logic.

As mentioned previously, I saw Gödel and his wife again in 1948, with the war years in between. However, it seemed as if we had seen each other the day before. We were staying at the Institute of Advanced Study, living in one of their little houses, attached to the von Neumann project. The war years had transformed us into numerical analysts and my husband Jack had close connections with von Neumann when the latter visited Great Britain. In fact Jack travelled with him to various Navy installations and even got him interested in high speed computing machines.

As in the old times I invited Gödel to tea, this time, of course, with wife, and to have it more of an Austrian party I also invited Oskar Morgenstern whom he knew very well, of course.

In 1951 Gödel gave his Gibbs lecture (Some basic theorems on the foundations of mathematics and their philosophical implications), one of the biggest honours of the American Mathematical Society. I sat with them and when Gödel returned Mrs. Gödel said: 'Kurtele, if I compare your lecture with the others, there is no comparison!' Although Kurt knew, of course, that his wife could not understand the lecture, a flattered smile appeared on his face!

Gödel had not yet obtained the title Professor in 1948. It is hard to believe. There was a time when Einstein's influence there was not strong. But I do not know whether Einstein tried to do something about it. But Einstein had great friendship for Gödel. I do not know who his other friends were apart from Morgenstern.

He was very sad about the death of the young logician Spector. Last year (1983) Mrs. Renée Robinson showed me a letter expressing Gödel's admiration for the work of Abraham Robinson. Apparently Gödel had hoped that Abby would be his successor at Princeton. Fate decided differently — for Abby died earlier.

APPENDIX

Ernst Zermelo - A letter to Reinhold Baer

Freiburg 7-10-31 - Karlstr. 60

Lieber Herr Baer, beifolgend ein Durchschlag meines *neuen* Berichtes über den Elster-Vortrag, der auf Wunsch für die «Forschungen und Fortschritte» hergestellt wurde, nach Möglichkeit «gemeinverständlich», aber schon wegen der Kürze (4 Seiten) gewiß *nicht leicht* verständlich. Der ist natürlich ganz *ohne Polemik* oder historische Bezugnahme, enthält *nur* Positives. Der andere (für die *Mathematische Vereinigung*) *soll und muß* Widerspruch erregen bei denen, die es trifft. Denn ich bin meiner Sache ganz sicher und wünsche durchaus, daß die Frage einmal in Fluß kommt. In Bad Elster habe ich noch im Vortrage selbst und nachher eine direkte Polemik gegen Gödel vermieden: regsame Anfänger soll man nicht abschrecken. Ich habe extra Gödels Vortrag *vor* dem meinigen ansetzen lassen und eine *gemeinsame* Diskussion beantragt. Aber die Folge meiner Loyalität war einzig die, daß die ganze Diskussion auf den unberechtigten Vorschlag Fraenkels (bei *jeder* Gelegenheit fällt er mir in den Rücken!) *noch* weiter zurückgestellt wurde und dann im Sande verlief.

Mein eigener Vortrag fiel dabei ganz unter den Tisch. Jetzt will ich es anders machen: die Herren *müssen* endlich Farbe bekennen, wenn ich öffentlich erkläre, daß der vielbewunderte Gödel'sche «Beweis» Unsinn ist; denn mit ihm fällt der ganze Skolemismus: insbesondere in der Form des Carnap'schen «PM-Systems». Die Sache ist doch ganz einfach. Gödel geht von folgenden Voraussetzungen aus. 1) *Annahme A,* daß alle Satzfunktionen $R(n, x)$ des Systems (Gödel schreibt dafür: $[R(n); x)]$ eine *abzählbare* Menge bilden, während auch die Variable x auf die Zahlenreihe beschränkt ist. Diese Annahme kommt also (verallgemeinert) darauf hinaus, daß die Menge der *Funktionen* dem Wertevorrat der *Variablen x ein-eindeutig zugeordnet* ist. 2) die *Annahme B*, daß man (analog dem Cantor'schen Diagonalverfahren) durch eine über *alle* Satzfunktionen erstreckbare Quantifikation *neue* Satzfunktionen *innerhalb des* Systems definieren könne. Diese *beiden Annhmen* stehen zueinander im Widerspruch, da nach Cantor die Menge der Funktionen von *höherer Mächtigkeit ist.* Gödel benutzt gleichzeitig *beide* Annahmen A, B und kann dann natürlich *alles* beweisen. Warum nur die Existenz «unentscheidbarer Sätze», warum nicht die «Unsterblichkeit der Maikäfer» oder gar die «Unfehlbarkeit der Fakultäten»? Auf meinen Brief hat er mir noch immer *nicht* geantwortet. Augenscheinlich hat er nichts mehr zu «meckern». Könnten Sie nicht einmal in Ihrem Kolloquium in Halle die Frage zur Sprache bringen, indem Sie etwa *gleichzeitig* über Gödels und meinen Vortrag referieren?

Die beiden Durchschläge können Sie behalten, bis ich sie einmal selbst brauche und um Rücksendung bitte; ich habe ja noch ein weiteres Exemplar. Leider kann ich selbst nicht kommen — bei der 12stündigen Reise und den hohen Fahrkosten.

In Berlin habe ich auch nichts mehr zu suchen und in Leipzig ebensowenig. Die Frage der Antinomie Richard und des Skolemismus *muß* endlich *ernsthaft* erörtert werden, nachdem leichtsinniger Dilettantismus wieder am Werke ist, das ganze Forschungsgebiet zu diskreditieren — wie Schoenflies und Genossen seinerzeit die Mengenlehre diskreditierten. In meinem «Fundierungsprinzip» glaube ich endlich die richtige Handhabe gefunden zu haben, um alles Erforderliche aufzuklären. Aber niemand hat's verstanden, wie auch noch niemand auf meine Fundamenta-Arbeit reagiert hat — nicht einmal meine guten Freunde in Warschau. Hilbert sagte freilich einmal: eine Arbeit brauche 15 Jahre, um gelesen zu werden. Nun Hilbert selbst hat ja mit seiner «Geometrie» schnelle Erfolge gehabt; aber war das auch wirklich eine große Leistung? — Das Fundierungsprinzip beherrscht *alle* möglichen Satzsysteme, auch das Russell-Carnap'sche, soweit es widerspruchsfrei ist. Die «finitistische» Annahme A) kommt auf eine *Beschränkung* der Quantifikationsstufe hinaus und ist dann natürlich mit einer «freien» Begriffsentwicklung im Sinne des Diagonalverfahrens B) *unvereinbar.* Also man entscheide sich für A) oder B): tertium non datur. Mit besten Grüßen an Sie und die Ihrigen

E. Zermelo

Was macht Frl. Kröncke? Ich habe schon lange nichts mehr von ihr gehört.

Left: Karl Menger, Kurt Gödel;
second from right: Olga Taussky-Todd

Ernst Zermelo
(1871-1953)

Freiburg, 7th October, 1931 - Karlstr. 60

Dear Mr. Baer,

Enclosed a copy of the *new* summary of my Bad Elster lecture, which was written for «Forschungen und Fortschritte» on request; so far as possible it is «generally understandable» but if only because of its brevity (4 pages) certainly *not easy* to understand. It is, of course, completely *free of polemics* or historical reference and contains nothing negative. The other account (for the *Mathematische Vereinigung) should and must* provoke contradiction on the part of those whom it affects. For I am very sure of my position and would greatly like to see the question taken up. At Bad Elster I avoided any direct polemic against Gödel both in the lecture itself and afterwards: one should not frighten off enterprising beginners. I deliberately had Gödel's lecture put on *before* mine and asked that they be discussed *together*. But the sole consequence of my loyalty was that the whole discussion was put *further* back at the unjustified suggestion of Fraenkel (who stabs me in the back at every opportunity!) and then petered out.

My own lecture was then left completely out of account. Now I am going to do it differently: the gentlemen will *have to* declare their hand finally when I publicly assert that Gödel's much-admired 'proof' is nonsense; for with it the whole Skolem doctrine collapses, especially in the form of Carnap's 'PM System'. The matter is, after all, quite simple. Gödel begins with the following presuppositions: 1) *Assumption (A):* all propositional functions $R(n, x)$ of the system (Gödel writes $[R(n), x]$) constitute a *denumerable* set, while also the variable x is restricted to the number series. This assumption, then, means (when generalized) that the set of the *functions* is in *one-to-one correspondence* with the domain of the *variable* x. 2) *Assumption (B):* one can (analogously to Cantor's Diagonal Procedure) define *new* propositional functions *within the* system by quantification over *all* propositional functions. These *two assumptions* conflict since according to Cantor the set of functions *is of greater power.* Gödel makes use of *both* assumptions, (A) and (B), at the same time and can then, of course, prove *anything.* Why only the existence of «undecidable sentences», why not the «immortality of the cockchafer» or even the «infallibility of the faculties»? He still has *not* answered my letter. Apparently he has nothing else to «grouse» about. Could you not raise the question for discussion sometime in your seminar at Halle by reporting on my lecture and Gödel's *at the same time*?

You can keep both copies till I need them myself and ask for them back — I've got a further copy. Unfortunately I cannot come myself — a 12 hour journey and the fares so high!

Anyway, I have no further business in Berlin, nor in Leipzig for that matter. The question of the antinomy of Richard and the Skolem doctrine *must* at last be discussed *seriously,* seeing that frivolous dilettantism is again at work to discredit the whole area of research — just as Schoenflies and comrades once discredited set theory. I believe I have at last found in my '*Fundierungsprinzip*' the right instrument for explaining whatever is in need of elucidation. But nobody understands it, just as nobody has yet reacted to my Fundamenta article — not even my good friends in Warsaw. After all, Hilbert once said it takes 15 years before a paper read. Now Hilbert himself had rapid success with his 'geometry', but was it really such a great achievement? The axiom of foundation governs all possible propositional systems, including that of Russell/Carnap, in so far as it is consistent. The «finitistic» assumption (A) amounts to a *restriction* of the level of quantification and is then, of course, *irreconcilable* with a «free» development of concepts in the sense of the Diagonal Procedure (B). So one must choose between (A) or (B); tertium non datur.

With best wishes to you and yours,
E. Zermelo

How is Miss Kröncke? I haven't heard anything from her in a long time.

Stephen C. Kleene

GÖDEL'S IMPRESSION ON STUDENTS OF LOGIC IN THE 1930s

Gödel completed his doctoral dissertation in the autumn of 1929. I completed mine in June 1933, less than four years later.

As I look back to the beginning fifty years ago of my career as a mathematical logician, it is natural for me to ask myself what influences, what impressions, what associations, contributed to determining the direction which it took.

So much depends, at the beginning of a research career, on the intellectual setting in which one is immersed. Shall I say that so much depends on there floating in it new ideas and concepts — ones with great potential implicit in them for developments not yet begun?

I am sure Gödel owed much to his participation in Hans Hahn's seminar and in Karl Menger's colloquium.

I was most fortunate (surely, whatever career I might otherwise have had, it would not have been at all the same) in being a student of Alonzo Church (beginning in the fall of 1931) and in being exposed to Gödel's thought, first in 1931 through his writings, and only a bit later, in the spring of 1934, personally. Their impressions on me were quite distinct, and I can say complementary.

Before taking Church's course in the fall term of 1931/32, my only acquaintance with mathematical logic and foundations had been very general, from reading Alfred North Whitehead's «An Introduction to Mathematics» *1911* and Bertrand Russell's «Introduction to Mathematical Philosophy» *1919*.

Church plunged right into the presentation of the new formal system introduced in his two papers, then in manuscript, entitled «A set of postulates for the foundation of logic» *(1932, 1933)*. This was strikingly novel, and as we now know two things were hidden below the surface: that the full system was inconsistent, and that there could be separated out from it a subsystem, the λ-calculus, the full potentialities of which, even after it was separated out, were initially quite unsuspected. I had a part in mining

GÖDEL REMEMBERED
Salzburg 10-12 July 1983
Ed. by P. Weingartner and L. Schmetterer
BIBLIOPOLIS 1987

these two hidden things, the first a disaster and the second a treasure. Above all, a student needs to stand on ground from which things can be dug.

Because of my sketchy previous knowledge of foundations, and Church's concentration virtually exclusively on his new formal system, I would have remained (how long I'm not sure) basically ignorant of the broad features of the landscape of logic and foundations of that day — of for example the classical propositional calculus, of the restricted functional (or first-order predicate) calculus, of formal number theory, of Hilbert's formalist program (other than in very general terms) — except that (and this is a big exception), a month or two after I had begun my study under Church, Gödel entered my intellectual life.

How? One day in the fall of 1931 the speaker in the mathematics colloquium at Princeton was John von Neumann. Instead of talking on work of his own (of which there was plenty), he spoke on the results of Gödel's *1931* paper, which had recently come out in the *Monatshefte*, but which Church and we in his course had not yet noticed. von Neumann had had a preview of the first of those results (accompanied by intellectual intercourse with Gödel) at the Königsberg meeting of September 1930.

After the colloquium, Church's course continued uninterruptedly concentrating on his formal system; but on the side we all read Gödel's paper, which to me opened up a whole new world of fascinating ideas and perspectives.

The impression this made on me was so much the greater because of the conciseness and incisiveness of Gödel's treatment. If I had been introduced to the world of foundations outside of Church's system by extensive reading in the other literature, the effect on me would have been less dramatic.

Actually, I first undertook such extensive reading, beginning with Hilbert-Ackermann *1928*, Hilbert-Bernays *1934* and Heyting *1934*, in the summer of 1936, as part of my preparation to teach a semester graduate course at Wisconsin. This course, called then «Mathematics 230» because it met at 2:30 in the afternoon, was perhaps the first course in the United States aiming to give a comprehensive treatment, with all the necessary preliminaries and related theory, reaching Gödel's theorems as the culmination of the course.

The course became essentially the first half of my book «Introduction to Metamathematics» *1952*. The eighth reprint of the book appeared in 1980, besides which it was translated into Russian, Chinese and Spanish, and received the supreme honor of being produced in Taiwan. The cover of the brochure «North-Holland Publications in Mathematics and Logic 1983» is adorned with an array of formulas photocopied from page 119 of that book. A number of copies were sold in Salzburg at a 25% discount. This all speaks very highly of the wide and enduring interest a treatment of foundations — of metamathematics — centered around Gödel's results has sustained over the years.

After completing my PhD thesis at Princeton in June 1933, I was away from Princeton for seven months (except for a visit of a few days to take my final PhD examination). On my return in February 1934, there was Gödel in the flesh. He was in residence at the Institute for Advanced Study for the academic year 1933-34.

Hardly had I got back to Princeton than it developed that Gödel would give an exposition of his undecidability results in a series of lectures entitled «On undecidable propositions of formal mathematical systems». I think it was Oswald Veblen who suggested that Rosser and I take notes on the lectures. We organized what we could have called «The Gödel Notes Company , Unincorporated», solicited subscriptions, and collected the fees for them. I have forgotten the price (maybe 2 or 3 dollars). The notes were distributed in installments at short intervals as the lectures progressed. The entire packet of 28 mimeographed pages, plus 2 pages of «notes and errata», received Gödel's blessing in May or June 1934. In an undated letter from the Hotel Edison, New York, where he stayed on the eve of his departure for Europe, he wrote me, «I think that now everything in my notes is stated very clearly. There are only some minor corrections, which I wrote in the text and on the sheets enclosed.»

Copies of the *1934* notes, besides being distributed to the subscribers, went to a number of libraries. They were cited in at least a dozen bibliographies (namely in Church's «A bibliography of symbolic logic» in the *Journal of Symbolic Logic,* vol. 1 (1936), and in books of Quine 1940, Church 1941, Kattsoff 1948, Kleene 1952, Fraenkel 1953, Péter 1957, Davis 1958, Fraenkel-Bar Hillel 1958, Curry 1963, Kneebone 1963 and Mendelson 1964), before being printed in Davis' «The Undecidable» *1965*. They will be reprinted in the forthcoming first volume of Gödel's collected works, which is in preparation by an editorial board (with Solomon Feferman as editor-in-chief) established by the Association for Symbolic Logic.

Of course, in this process of note taking, with Gödel editing the notes, there was an intimacy in the communication between him and us, his students, not existing in our earlier exposure to his thought through reading his *1931* paper.

And after being put onto that paper by von Neumann's colloquium lecture, our attention was drawn to his *1930* paper on the predicate calculus,which treats the completeness problem for that in an incisive way (with a new proof of Löwenheim's *1915* theorem, and with what is now called «compactness» being established). It had cried out for treatment thus, since Hilbert and Ackermann in *1928* had formulated the problem (apparently unaware of the solution of it that was implicit, but not pinned down formally, in the *1922* paper of Skolem, who also seemed to be unaware of this consequence of his work). Indeed, Gödel had a style, which greatly impressed us as students in our formative years.

Subsequently to 1934, I had personal contacts with Gödel during the two years I was a visitor at the Institute for Advanced Study (1939-40, 1965-66) and the year I was a visiting professor at Princeton (1956-57). My parents and I had the pleasure of two visits of several days each by Gödel and his wife Adele to our farm in Maine in 1941.

How did Gödel, the man and his work, affect our careers?

I came in contact with simple examples of recursions in Church's course. But I learned an exact characterization of the class of the «primitive recursive functions» (as they have been called since my *1936* paper on general recursive functions), and of various of its closure properties, from Gödel *1931.* True, I could have found much of this in Skolem *1923* and Hilbert *1926*, had I known where to look.

I alluded at the beginning of this talk to the λ-calculus as being separated out from the full formal system of Church *1932*, *1933*. This has been of interest to logicians and computer scientists over the intervening half century. Witness the 1981 book of H.P. Barendregt (Utrecht) «The Lambda Calculus. Its Syntax and Semantics» (North-Holland Publishing Co.), 615 pp. Also the Proefschriften (Utrecht) of J.W.Klop «Combinatory Reduction Systems» Amsterdam 1980 (323 pp.) and Adrian Rezus «Lambda-Conversion and Logic» Utrecht 1981 (197 pp.)

It was between February 1932 and the completion of my PhD thesis in June 1933 (published in a revised version in *1935*) that we learned the extraordinary potentialities of the λ-calculus.

In a letter to Paul Bernays dated January 23, 1935, Church wrote, «The most important results of Kleene's thesis concern the problem of finding a formula to represent a given intuitively defined function of positive integers (it is required that the formula shall contain no other symbol than λ, variables, and parentheses). The results of Kleene are so general and the possibilities of extending them apparently so unlimited that one is led to conjecture that a formula can be found to represent [i.e., as we now say, to λ-define] any particular constructively defined function of positive integers whatever.»

Church had proposed this conjecture to various logicians at Princeton early in 1934, including Gödel. (The term «λ-definable» came in with Church's *1936* and my *1936a.*) Gödel was unpersuaded.

In a letter to me of November 29, 1935, Church, referring to discussions with Gödel of around February or March 1934, wrote, «My proposal that lambda-definability be taken as a definition of [effective calculability] he regarded as thoroughly unsatisfactory. I replied that if he would propose any definition of effective calculability which seemed even partially satisfactory I would undertake to prove that it was included in lambda-definability. Evidently it occurred to him later that Herbrand's definition of recursiveness [proposed in a 1931 letter of Herbrand to Gödel], which had no regard to effective calculability, could be modified in the direction of effective calculability, and he made this proposal in his lectures. [It is in the last section of the *1934* lecture notes.] At that time he did specifically raise the question of the connection between recursiveness in this new sense [what we now call «general recursiveness»] and effective calculability, but said that he did not think the two ideas could be satisfactorily identified 'except heuristically'.»

Thus in those years Gödel distanced himself from Church's identification published in *1936* of *the λ-definable functions*, and of *the general recursive functions* (for which two classes of functions I had by June 1935 a proof that they are the same, published in my *1936a* with *1936*), with the class of the functions intuitively recognized as «effectively calculable». This came to be called «Church's thesis» with my *1952* and almost with my *1943*.

Here «effectively calculable» (in the terminology of Church *1936* and of his November 29, 1935 letter to me), or «constructively defined» (in the terminology of his January 23, 1935 letter to Bernays), is also expressed by «calculable by an algorithm». (The term «algorithm» is a contraction of the last part of the name of the ninth century Arabian mathematician, Abu Abdullah abu Jafar Muhammad ibn Musa al-Khowarizmi.) Over two thousand years of mathematical history are studded with examples of procedures mathematicians have accepted as being «algorithms», but without there having been a precise formulation in general of what an «algorithm» is, say for the case of «algorithms» for calculating number-theoretic functions. This indicates the profound significance of Church's thesis. The thesis first provided a conceptual basis for attempting (successfully as it proved) to give negative solutions to the Entscheidungsproblem for certain formal systems (showing that no algorithm exists for deciding which formulas are provable in them), like those of first-order logic and of formal number theory. The Entscheidungsproblem for various formal systems had been posed by Schröder *1895*, Löwenheim *1915* and Hilbert *1918*. The negative solution for the case of first-order logic was given by Church in *1936a* (using his thesis) and by Turing in *1937* (with his equivalent form of the thesis), and for formal number theory by Church in *1936* and later by Rosser in *1936* with a weakening of the hypotheses on the system.

Gödel later accepted Church's thesis, which had become the Church-Turing thesis on the basis of Turing's *1937*, which gave a third characterization of the same class of functions — what are now called *the Turing computable functions*. It must be mentioned, though Gödel did not so far as I know ever cite it, that Emil Post in *1936* (received October 7, 1936) independently of Turing originated essentially the same idea as Turing introduced in *1937* (received May 25, 1936).

But while Gödel stood aside at the time from giving to his general recursive functions the absolute significance which Church (and those, like me, who had become converts to his thesis) ascribed to them, he had provided us with the notion of them, which we exploited in developing the theory of the class of the functions in question.

Thus, beginning with my paper *1936*, I mainly worked with the version of the theory of the effectively calculable number-theoretic functions using Gödel's general recursive functions. In *1936*, I proved a normal form theorem for the functions of the class considered. Maybe I might have thought of this in terms of the λ-definable functions. But in fact it came to me as a useful lemma (the lemma being in retrospect at least as interesting as the theorem) for my proof in *1936a* that every general recursive function is λ-definable.

In that *1936* paper I reformulated Gödel's definition of general recursiveness, just a bit as it seemed to me. In conversation with me on October 31, 1979, Martin Davis expressed the opinion that the equivalence between Gödel's definition of general recursiveness and mine (which equivalence Gödel, in a February 15, 1965 letter to Davis, called «not quite trivial»), and my normal form theorem, were considerations that combined with Turing's arguments to convince Gödel of the Church-Turing thesis.

As Church wrote me on November 15, 1935, «Both Gödel and Bernays are very much interested in your result that all general recursive functions can be obtained from primitive recursive functions by means of the epsilon operator [the least-number operator]. It is, of course, natural that they should be quicker to see the importance of a result like this than of one concerning lambda-definition or the like.»

To put it colloquially, general recursiveness had at that time more «sex appeal» than λ-definability, though the 1135 pages of 1980-1981 publications on λ-definability (and the related combinatory reducibility) prove that λ-definability did have quite a lasting charm.

Another project which appealed to me then was to put Gödel's famous first incompleteness theorem,which in *1931* was for «Principia Mathematica and [quite a broad class of] related systems», into a general context as resting simply on the effectiveness or mechanical character of proof in a formal system, which under Hilbert's formalist program is the whole purpose of having formal systems, plus in addition only the requirement that the system embody correctly a modicum of informal number theory.Thus in *1936*, and in a simpler way in *1943*, I formulated generalized versions of Gödel's first incompleteness theorem, indeed in *1943* fitting it into a setting in a new theory of a hierarchy, of the arithmetical predicates.

In *1938*, I generalized (or should I say «partialized»?) Gödel's notion of «general recursive functions» to get the notion of «partial recursive functions».

It may amuse you that, in conversation with Gödel in the fall of 1939, I mentioned «partial recursive functions»; and he came right back at me with the question, «What is a partial recursive function?». He evidently had not looked at my *1938* paper.

Apparently, he then embraced the idea. In conversation with him in the summer of 1941, I told him about my «realizability» interpretation of intuitionistic number theory, using partial recursive functions. He then told me that he had another interpretation, also using partial recursive functions. I do not remember the details. So I do not know how it was related to that in his *1958 Dialectica* paper, on which he lectured at Princeton and Yale in 1942 (Wang *1981* p. 657).[1]

As a further illustration of Gödel's approving my idea of «partial recursive functions», Hao Wang in his *1974* book reports on p. 84 as follows: «Gödel points out that the precise notion of mechanical procedures is brought out clearly by Turing machines producing partial rather than general recursive functions.»

I did not have close relations with Gödel over the years, unlike some logicians who had frequent discussions with him and recorded the results. What I took from him came mainly from his writings. And I have been able to report that the ideas and techniques in them stimulated me to do work that I would not otherwise have done, and on occasions to report something new back to Gödel.

A survey lecture to be entitled «The work of Kurt Gödel» was scheduled to be given at the meeting of the Association for Symbolic Logic on December 29, 1975 in New York City. Hartley Rogers, then the President of the Association, invited me to give the lecture. I initially declined, because of the pressure of other work. But Rogers then told me that Gödel had been informed that there was to be such a lecture, and had requested that I be the lecturer. Thereupon I felt obliged to accept. Gödel did not attend. The lecture was published in *1976*. I had slipped on one detail, to which Gödel drew my attention through an intermediary, so that I published an addendum in *1978*. In *1976* I gave some space to my *1943* generalized version of his first incompleteness theorem, which was elaborated in my *1952* book. To my knowledge he never commented on that *1943* and *1952* writing of mine, though in a «*Note added* 28 *August* 1963» to the translation of his *1931* in van Heijenoort *1967,* and in the «Postscriptum... June 3, 1964» to the Davis *1965* printing of his *1934* notes,

[1] In a correction to his *1983*, Dawson says that the Yale lecture was on April 15, 1941. There is a reference, in Section 5 of Kreisel's lecture at this Symposium, to material in Gödel's Nachlass from the 1940's bearing on his functional interpretation. Unfortunately, I cannot provide any information in response to the three queries addressed to me in Sections 5 and 6.

he lent his approval to a generalized version of his incompleteness theorems. He did not touch on the fact, remarked in my *1943* and *1976* papers, that the undecidable (indeed, true but unprovable) propositions for all formal systems embracing correctly a certain amount of formalized elementary number theory can be taken to be values of one preassigned number-theoretic predicate, e.g. the predicate $(\overline{Ey})T_1(x, x, y)$ (equivalently $(y)\bar{T}_1(x, x, y)$) first introduced in my *1936*. But I take his silence on this in the exchange which led to my *1978* addendum, and even more his having chosen me to be the lecturer in 1975, to witness his having looked with favor on this work of mine done in response to the impression his original work had made on me. He could not have been unaware of the extensive writing of mine along these lines in my *1952* book, to which he made a reference (though not to that part of it specifically) on p. 73 of the Postscriptum in Davis *1965*.

So it came about that I was over the years a practitioner of recursion theory, continuing along a path onto which impressions received from Church and from Gödel had steered me. I was not alone. Thus, a few months over a half century since I came under their distinct but complementary impressions, the American Mathematical Society's Thirtieth Summer Research Institute was on Recursion Theory, at Cornell University, June 28 - July 16, 1982, with contributions by (at least) 73 investigators.[2] Recursion theory had not remained a small operation.

J. Barkley Rosser was the other student in Church's graduate course in the fall term of 1931/32 who remained active as a logician over the succeeding half century.

Like me he began with research that grew out of Church's project. Specifically, in his thesis *1935* he established a relationship between Church's λ-calculus and the combinatory logic, which had originated with Schönfinkel in *1924* and been further developed by Curry in *1929, 1930* and *1932*. This relationship was of fundamental importance (as I applied it in *1934*) to the development of the potentialities of λ-definability. He joined with me in *1935* in establishing the inconsistency of Church's full

[2] The magic number 73 is the sum of the number of the names of authors or coauthors of notes distributed by the Institute which I possess and the number of the names (e.g. «Kleene») of other persons listed as speakers on program sheets distributed by the Institute w.i.p.

system (and of a system considered by Curry), and with Church in *1936* in proving a fundamental consistency property of the λ-calculus.

But then, like me, he did work which was a result also of the impression Gödel had made on him. This led to three papers of his of which I now quote the titles: «Extensions of some theorems of Gödel and Church» *1936*; «Gödel theorems for non-constructive logics» *1937;* and «An informal exposition of proofs of Gödel's theorems and Church's theorem» *1939.* Incidentally, his *1937* footnote 11 contains the only published reference to another piece of research I did as a consequence of Gödel's impression.

Alan M. Turing, after writing at Cambridge, England his fundamental *1937* paper on his computability notion, came to Princeton at the end of September 1936 to work for a PhD under Church. He took as his topic one growing out of Gödel's famous *1931* first undecidability theorem, but treated using the λ-calculus. Thus Turing's *1939* combined in one work the impressions on him of Church and Gödel, while his *1937* paper (which in my view was done independently of Gödel's work) greatly impressed Gödel.

Gödel, in his position at the Institute for Advanced Study, was not circumstanced to supervise PhD students. But his influence on the three PhD students of Church whom I have named, if not while they were working for their PhD's only shortly after, was not less than a thesis supervisor ordinarily has on his students.

Gerhard Gentzen's famous *1934-5* paper on investigations concerning logical deduction was his inaugural dissertation at Göttingen. But clearly Gentzen's idea of attacking Hilbert's consistency problem for the case of pure number theory in his *1936* paper by using a new finitary method (ε_0-induction) reflects an impression on him by Gödel through the second incompleteness theorem of Gödel *1931.* Further writings of Gentzen which reflect the impression of Gödel on him are his papers «Die gegenwärtige Lage in der mathematischen Grundlagenforschung» *1938* and «Neue Fassung des Widerspruchsfreiheitsbeweises für die reine Zahlentheorie» *1938a.*

Had I not confined myself in this talk to Gödel's impression on students of logic in the 1930s while or shortly after they were PhD students, I could not possibly have related in the allotted time all the manifestations of Gödel's impression on students of logic.

REFERENCES

A date shown in the text in italics accompanying mention of a person constitutes a reference to a work listed here.

CHURCH, ALONZO

1932. A set of postulates for the foundation of logic. *Annals of Math. 2s 33*, 346-366.
1933. A set of postulates for the foundation of logic (second paper). Ibid. *34*, 839-864.
1936. An unsolvable problem of elementary number theory. *Amer. J. Math. 58*, 345-363.
1936a. A note on the Entscheidungsproblem. *J. Symbolic Logic 1*, 40-41.Correction, 101-102.

CHURCH, ALONZO AND ROSSER, J.B.

1936. Some properties of conversion. *Trans. Amer. Math. Soc. 39*, 472-482.

CURRY, HASKELL B.

1929. An analysis of logical substitution. *Amer. J. Math 51*, 363-384.
1930. Grundlagen der kombinatorischen Logik. Ibid. *52*, 509-536, 789-834.
1932. Some additions to the theory of combinators. Ibid. *54*, 551-558.

DAVIS, MARTIN

1965. *The Undecidable. Basic Papers on Undecidable Propositions, Unsolvable Problems and Computable Functions*. Hewlett, N.Y. (Raven Press), v+440 pp.

DAWSON, JOHN W., JR.

1983. The published work of Kurt Gödel: an annotated bibliography. *Notre Dame J. Formal Logic 24*, 255-284. Addenda and Corrigenda forthcoming.

GENTZEN, GERHARD

1934-5. Untersuchungen über das logische Schliessen. *Math. Zeitsch. 39*, 176-210, 405-431.

1936. Die Widerspruchsfreiheit der reinen Zahlentheorie. *Math. Annalen 112,* 493-565.
1938. Die gegenwärtige Lage in der mathematischen Grundlagenforschung. *Forschungen zur Logik und zur Grundlegung der exakten Wissenschaften n. s. no. 4,* Leipzig (Hirzel), 5-18.
1938a. Neue Fassung des Widerspruchsfreiheitsbeweises für die reine Zahlentheorie. Ibid., 19-44.

GÖDEL, KURT

1930. Die Vollständigkeit der Axiome des logischen Funktionenkalküls. *Monatsh. Math. Phys. 37*, 349-360.
1931. Über formal unentscheidbare Sätze der Principia Mathematica und verwandter Systeme I. Ibid. *38*, 173-198
1934. On undecidable propositions of formal mathematical systems. Mimeographed, Princeton, N.J., 30 pp. Printed in Davis *1965*, 39-74.
1958. Über eine bisher noch nicht benützte Erweiterung des finiten Standpunktes. *Dialectica 12,* 280-287.

HEYTING, AREND

1934. Mathematische Grundlagenforschung: Intuitionismus - Beweistheorie. *Ergebnisse d. Math. u. ihrer Grenzgebiete, vol. 3, no. 4.* Berlin (Springer), iv+73 pp.

HILBERT, DAVID

1918. Axiomatisches Denken. *Math. Annalen 78,* 405-415.
1926. Über das Unendliche. Ibid. *95*, 161-190.

HILBERT, DAVID AND ACKERMANN, WILHELM

1928. Grundzüge der theoretischen Logik. Berlin (Springer), viii+120 pp.

HILBERT, DAVID AND BERNAYS, PAUL

1934. Grundlagen der Mathematik, vol. 1. Berlin (Springer), xii+471 pp.

KLEENE, STEPHEN C.

1934. Proof by cases in formal logic. *Annals of Math. 2s. 35, 529-544.*
1935. A theory of positive integers in formal logic. *Amer J. Math. 57*, 153-173, 219-244.
1936. General recursive functions of natural numbers. *Math. Annalen 112*, 727-742.
1936a. λ-definability and recursiveness. *Duke Math. J. 2, 340-353.*
1938. On notation for ordinal numbers. *J. Symbolic Logic 3, 150-155.*
1943. Recursive predicates and quantifiers. *Trans. Amer. Math. Soc. 53,* 41-73.
1952. Introduction to Metamathematics. Amsterdam (North-Holland Pub. Co.), xi+550 pp. Eighth reprint, 1980.
1976. The work of Kurt Gödel. *J. Symbolic Logic 41*, 761-778
1978. An addendum. Ibid. *43*, 613.

KLEENE, S.C. AND ROSSER, J.B.
1935. The inconsistency of certain formal logics. *Annals of Math. 2s. 36*, 630-636.

LÖWENHEIM, LEOPOLD

1915. Über Möglichkeiten im Relativkalkül. *Math. Annalen 76*, 447-470.

POST, EMIL L.

1936. Finite combinatory processes - formulation I. *J. Symbolic Logic 1*, 103-105.

ROSSER, J. BARKLEY

1935. A mathematical logic without variables. *Annals of Math. 2s. 36*, 127-150 and *Duke Math J. 1*, 328-355.
1936. Extensions of some theorems of Gödel and Church. *J. Symbolic Logic 1*, 87-91.
1937. Gödel theorems for non-constructive logics. Ibid. *2*, 129-137.
1939. An informal exposition of proofs of Gödel's theorems and Church's theorem. Ibid. *4*, 53-60.

RUSSELL, BERTRAND

1919. *Introduction to Mathematical Philosophy*. London (Geo. Allen and Unwin), New York (Macmillan), viii+208 pp.

SCHÖNFINKEL, MOSES

1924. Über die Bausteine der mathematischen Logik. *Math. Annalen 92*, 305-316.

SCHRÖDER, ERNST

1895. *Vorlesungen über die Algebra der Logik (exakte Logik). Vol. 3 Algebra und Logik der Relative part 1*. Leipzig (Teubner), viii+649 pp.

SKOLEM, THORALF

1922. Einige Bemerkungen zur axiomatischen Begründung der Mengenlehre. *Wissenschaftliche Vorträge gehalten auf dem fünften Kongress der Skandinavischen Mathematiker in Helsingfors vom 4. bis 7. Juli 1922*, Helsingfors 1923, 217-232.
1923. Begründung der elementaren Arithmetik durch die rekurrierende Denkweise ohne Anwendung scheinbarer Veränderlichen mit unendlichem Ausdehnungsbereich. *Skrifter utgit av Videnskapsselskapet i Kristiania, I. Matematisk-naturvidenskabelig Klasse no. 6*, 38 pp.

TURING, ALAN M.

1937. On computable numbers, with an application to the Entscheidungsproblem. *Proc. London Math Soc. 2s. 42*, 230-265. A correction, *43* (1937), 544-546.
1939. Systems of logic based on ordinals. Ibid. *45*, 161-228.

VAN HEIJENOORT, JEAN

1967. From Frege to Gödel: a Source Book in Mathematical Logic, 1879-1931. Cambridge, Mass. (Harvard Univ. Press), xi+660 pp.

WANG, HAO

1974. From Mathematics to Philosophy. London (Routledge & Kegan Paul), New York (Humanities Press), xiv+428 pp.
1981. Some facts about Kurt Gödel. *J. Symbolic Logic 46*, 653-659.

WHITEHEAD, ALFRED NORTH

1911. An Introduction to Mathematics. London (Williams and Norgate), New York (Henry Holt), vi+256 pp.

Georg Kreisel

GÖDEL'S EXCURSIONS INTO INTUITIONISTIC LOGIC

TABLE OF CONTENTS

(i) Improved formulation of the second incompleteness theorem: $\square(\square p \rightarrow p)$, valid for g.p., fails for formal derivability, where
(ii) an opposite extreme holds (Löb's theorem): $\square(\square p \rightarrow p) \rightarrow \square p$;
(iii) by (ii) the modal language distinguishes between formal and g.p., but not between g.p. and truth;
(iv) the interpretation of $\square$ as formal derivability provides a coherent mixture of object and metalanguage.

(c) *Infinitely many* — and hence new — *monadic propositional operators* are found in (6) on limitations of finite truth tables for i.l.

(i) Examples of positive and negative results on new propositional fragments of i.l.;
(ii) contrast of (i) with the singular functional completeness of truth functional propositional logic, and parallels to abstract model theory;
(iii) a parallel between the untutored notions of number and of proposition adumbrates two philosophical lessons on
the role of new operators when i.l. is interpreted as concerning a larger class of propositions (than thought of in classical logic) and on i.l. considered as a formal generalization of classical logic;
with a remark on prospects for theories of natural languages in (d) below.

(d) *Natural language,* and the style of current linguistic studies are viewed in the light of experience with i.l., by reference to the following items:

(i) relevance of mathematical language and grammatical rules in terms of survival value, not only frequency;
(ii) universal semantical scheme as opposed to a relatively small arsenal of meanings for many situations, and the skills needed for these alternatives;
(iii) the thresholds of sophistication in the alternatives of (ii);
(iv) selection of situations rewarding theoretical studies by anything like currently available means, and
(iv) some implications of the neglected distinction between natural and logical sense (when combining propositions with either sense).
(v) Natural history resembles — in aims and methods — the current style of studying natural language, in contrast to science, which differs from natural history less in the literary form of a mathematical presentation than the selection of phenomena, touched in (iv) above.

3. Effective rules

with distinctions between — idea(lization)s of — the systems which the rules are to affect, were a life-long interest of Gödel, whose own contributions range between opposite extremes. (a) and (b) concern primarily *given* languages for the rules:

(a) *equational rules,* with attention to the choice of calculus demanded by informal rigour, and limited scientific value of any such choices with corollaries on the scientific value of informal rigour (relevant to philosophy in the sense of the manifesto)

(b) *extension of the equational language* to more general logical languages in the notion of *formal computability,* which goes beyond earlier work by
(i) using formal rules for generating, and not only checking computations or derivations, and
(ii) being invariant or 'absolute' for wide classes of extensions of suitable starting systems.

(c) *Effectiveness for the perfect mathematician* is unquestionably the meaning of 'effective' implicit in foundational debates; with Turing's convincing analysis of effectiveness for the perfect computer providing the single most useful concept for describing — similarities and contrasts with — the perfect mathematician.
(i) The intuitionistic version of the perfect mathematician, known as 'creative subject', serves as foil to
(ii) Gödel's own version that is allowed to operate also on notions not conceived of as having been, in turn, created by the creative subject, and by-products of (i) and (ii) with impeccable scientific or technical interest are touched.

4. Effective rules of finite types

is an anecdotal presentation of (13), the socalled Dialectica interpretation, with antecendents and sequels

(a) *Gödel's own account in Oct. 55 of the early background:*
(i) shortcomings w.r.t. the kind of E-theorems wanted, and
(ii) in particular, w.r.t. some of Kleene's realizabilities;
(iii) digression on neglected advantage w.r.t. information about provably total recursive functions, and defects of partial functions used in realizability;
(iv) comparison with Kleene's realizabilities for negative formulae;
(v) loose ends about interpreting terms of higher type, in particular, as computation rules.

(b) More background on *Gödel's scepticism in 1955 about logic* as a tool for foundations or philosophy (as he saw these subjects in contrast to the view of section 1)

(c) *Principal progress* during the couple of years before Gödel wrote (13), especially with interpreting terms of higher type differently from a(v) above;

(i) Hereditarily effective and continuous operations, and methods needed to prove their closure under primitive recursion;
(ii) Extension of the language of first order arithmetic (used by Gödel) to second and all finite orders to avoid apparent absurdities; recursiveness of functional interpretations of all *true* formulae in negative fragments;
(iii) inadequacies: bounded types, HEO for higher order language, recursiveness in the original sense of S1-S9 (when defined on the full hierarchy, but not — as observed recently — on the hierarchy of continuous functions);
(iv) obvious sharpening of (ii) for theorems of *formal* systems, and choices that are logically needed in current mathematics; comparison with the no-counterexample interpretation.
(v) Omission of hereditarily recursive operations (HRO) without extensionality.

(d) *Gödel's last full-fledged paper* — (13), for Bernays' 70th birthday — is seen to fit Hilbert's ideas 30 years earlier, but with a twist: a primitive notion of function of higher type;
(i) asymmetry between rules and the ranges of their arguments, with reminders of familiar models,
(ii) definitional and demonstrable equality between terms of higher type,
(iii) the fan theorem functional
Low marginal utility of the primitive notion, comparable to the case of the primitive notion subset-of in recent years (but not earlier on).

(e) *A sequel to* (13): (Brouwer's) *bar recursion of lowest type* extended (by Spector) to *all finite types:* candidates for significant differences.
(i) Relation between Brouwer's socalled fully analysed proofs of Π^1_1 theorems (without detour via complicated terms) and cut-free proofs(without detour via logically complicated formulae): bearing on extended bar induction; with a digression on the evidence of abstract principles and the kind of formal differences between their formalizations measured by proof-theoretic strength;
(ii) (Howard's) neglected alternative to Spector's proof
(iii) (Gödel's) oversight concerning the roles of higher types in set theory with the power set construction and in constructive function spaces.

(f) Gödel's last minute attempts to use (13) for aims of traditional philosophy (in notes for an English translation):
(i) Kant's notion of analytic axiom, and its difference from analytic proof in the mathematical tradition of purity of method,
(ii) neglected difference between demonstrable and definitional equality

Kurt Gödel

Kurt Gödel

(g) At an opposite extreme to (f), *a sober view of* (13) as providing
(i) a glimpse of the potential and limitations of higher types,
(ii) a formally independent Π_2^0 statement, comparably relevant in the domain of computation as soundness in metamathematics or (Gentzen's) ε_0-induction in the area of infinite descent,
(iii) logical dampers on sterile excitement for or against definitional equality.

APPENDICES

1. Gödel's expository style in the thirties:

a neglected aspect is the *increasing* emphasis on aims requiring constructive metamathematics.

(a) *Completeness of predicate logic* is least affected (involving some constructive paraphrase of 'logical validity')
(i) the most popular paraphrase around 1930 implied decidability; neglected were
(ii) paraphrases for constructivity in terms of recursion-theoretic definability of — suitable data for — models, and
(iii) analysis of completeness proofs in terms of (ii).

(b) The *incompleteness paper* and especially its supplements (cf. 2b) pay more attention to constructivist aims, but leave gaps related to
(i) the notorious weakness of negation in i.l., in particular,
(ii) the formal verification by intuitionistic metamathematics that Gödel's particular sentences are not refutable in ω- or 1 - consistent systems.
(iii) In a digression an up-to-date summary is given of improvements in formulating incompleteness (in the style of 4g concerning the Dialectica paper).
(iv) In another digression, a fall out of the Salzburg meeting, the Chinese remainder theorem, used only incidentally in the incompleteness paper, is shown to be quite central to Gödel's early mathematical interests, which included class field theory.

(c) *Relative consistency,* mentioned in the title of the monograph (10), and related, but different aims:

(i) relative consistency and conservation,
(ii) watering down interpretations in the sense of Tarski (Mostowski-Robinson) to relative consistency, and
(iii) sharpening the latter quantitatively in terms of the relative lengths of hypothetical proofs of inconsistency in the systems considered;

(iv) in terms of (iii), H. Friedman's converse to the analogue of (ii). In conclusion, the emphasis on finitist proofs of relative consistency is a philosophical error in the sense of section 1.

2. Some corresponding high spots of early classical and i.l., where 'early' means the periods 1880-1930, resp. 1930-1960.

(a) *Completeness w.r.t. intended meanings* came at the *end* of the periods considered. Whatever the causes — such as dubious doubts about the precision of those meanings — the very aim of completeness is quite abstract, meanings and formal rules being rarely relevant to the same context.

(i) *Scientific* frivolity of studying intended meanings without testing their relevance for — and the scientific value of — their intended purposes, and *philosophical* interest of settling questions (about such meanings) without close analysis; key word: basis theorems.
(ii) Examples of problems about meaning where corresponding rules are discovered to be useful, and vice versa.

(b) Some *early concocted meanings* of classical and intuitionistic systems in constructive, resp. set-theoretic terms.

(i) For classical predicate logic: Skolem's validity for arithmetically defined — as opposed to arbitrary — structures, and Herbrand's *champs finis*,
(ii) for i.l., Tarski's calculus of systems and topological interpretation, and Kleene's realizabilities.

(c) *Philosophical assessment of intended and concocted meanings* after general reminders on the inadequacy of logical languages for expressing central properties of these meanings: for example, power set construction for the set-theoretic meaning, structural properties of (fully analysed) proofs for the intuitionistic meaning. — Useful relations (not conflicts!) between concocted and intended meanings:

(i) Herbrand's theorem is used for the ordinary classical meaning, and
(ii) concocted (topological) interpretations are used for proving the completeness of i.l. for the intended intuitionistic meaning.
(iii) Speculations on using an arsenal of different meanings for solving such problems as: Is P = NP?

(d) *Warnings* against using (a) — (c) for history or methodology generally, and in particular against painstaking formal documentation out of all proportion to general understanding. Principal analogue: history of the planets, in particular,

(i) *outward* behavior, that is, planetary motion,
(ii) *inner* life, that is, chemical evolution.

NOTES

1. Intuitionistic rhetoric: objective versus subjective (in contrast to section 1 on i.l. and natural mathematical language).

(a) Samples of diverse *reactions to the rhetoric,*

(b) *Gödel's overt reactions,* especially around 1930, to Brouwer's (and Hilbert's not too different) rhetoric:
(i) general doubts about foundational debates,
(ii) different relevance of his theorems to (different) rhetorical flourishes,
(iii) changes during Gödel's life time,
(iv) constant temperamental incompatibility with Brouwer's style.

(c) *Related reminiscences of Brouwer.*

2. Disjunction and existence properties, with reminders of widespread misunderstandings.

(a) Speculations on *Gödel's early recognition* of these misunderstandings;

(b) reports on his interest around 1940 in the *E-property for Heyting's arithmetic*: the Dialectica interpretation versus realizability;

(c) *anecdote* from the 70's.

3. Two unpopular traditions in the pursuit of knowledge are (a) Gödel's favourite, and (b) natural history; cf. 2 d.

(a) The exploration of *naive ideas* is also called 'wisdom of the ancients':
(i) Newton was attracted by the widsom of the ancients.
(ii) Contemporary objections are often less compelling than the tradition itself, comparable to the case of — the tradition of — informal rigour.
(iii) An anecdote concerning demons.
(iv) With obvious exceptions the law of diminishing returns applies very soon to the elaboration of naive ideas.

(b) *Natural history* is a minority view competing with the wisdom of the ancients.
(i) Supplements to 2 d (vi) on linguistics
(ii) Some off-hand remarks by Gödel on computer languages

4. EFFECTIVE RULES: supplements to section 3.

(a) *Rules*, that is, initial conditions or 'settings' *for physical systems* or, more pedantically, for their current theories:

(i) mathematical and (ii) didactic contributions of [PR]

(b) I.l. regarded as a theory about *rules effective for the perfect mathematician;* supplements to 3 c:

(i) formal systems with the E-property cannot refute CT, short for 'Church's Thesis', and the *petitio principii* involved in restriction to formal systems when examining CT;
(ii) the idea of a creative subject proceeding in ω steps is contrasted with the relentless pursuit of knowledge along
(iii) various kinds of transfinite progressions.

(c) *Turing's ideas* on establishing limits of the *perfect mathematician* (p.m.) in terms of his notion of perfect computer:

(i) a(nother) *petitio principii* involved in using only formal theories of the p.m.; cf. b (i) above,
(ii) neglect of growth of p.m.,
(iii) ignorance of laws of growth of p.m.

5. THE LAST SENTENCES of several papers are defective. (Some of them express Gödel's last-minute afterthoughts.)

(a) $\forall^n\exists^2\forall^n$ *with equality;*

(b) *compactness theorem for first order logic.*

(i) restriction to countable sets in contrast to (3) on propositional logic, and
(ii) neglect of relevance for consequence.

(c) The 'axiom' V = L *as a 'completion'* of formal set theory

(d) Interpreting the *fan theorem* by use of the primitive notion of effective rule of finite type (central to (13); cf. 4 d).

(i) reminders on defined notions of higher type operations;
(ii) Gödel's view on the role of his last sentence in (13) for Spector's [S], and
(iii) his proposals for formulating that view.
Finally, there are two digressions on
(iv) Spector's actual background, and
(v) intended and discovered ambiguities in Gödel's conversations and writings.

6. SOME CONVERSATIONS ON THE PROPER ORDER OF PRIORITY in logical research

(a) Formal results by *inspection of informal notions*; as in Gödel's work on L, and in contrast to Cohen's flexible method of constructing models, but also in contrast to

(b) *Gödel's own style in the forties,* for proving AC formally independent; more generally,

(i) contrasts between philosophical 'positions' and logical practice (Gödel and Cohen)
(ii) musings on long-term and short-term effects of ad hoc solutions for fruitful problems.

(c) Pertinent instances of the difference, stressed throughout this article, between *Gödel's early and later styles.*

7. FURTHER CONVERSATIONS ON TITLES, terminology and other devices for communicating (logical) knowledge.

(a) The slogan: *the meaning of a theorem is its proof*

(b) Some *criticisms by Gödel* of (my) clumsy choices of titles and terminology, in particular,

(i) about arithmetically defined models of set theory, providing so far undecided Δ_2^0 theorems, and model-theoretic proofs of the second incompleteness theorem,
(ii) basis theorems, and (iii) lawless sequences.

(c) Musings on the *manipulation of readers,*
preceded by a digression on survival of the fittest terminology:
(i) a role of formal details in the incompleteness paper,
(ii) a potential use for a part II of that paper.

(d) Digression on *Gödel's appreciation of logical work* in styles very *different from his own* (besides cases already considered in earlier notes),
(i) Gentzen, (ii) Kleene, (iii) Dana Scott, (iv) Takeuti.

8. AFTERTHOUGHTS: several should have been integrated in the text or in earlier notes

(a) *Solemn questions on musings* about logical research.
(i) Notes 3 and 6: working harder on the wisdom of the ancients
(ii) See-saw concerning the strategy in Note 6, with particular attention to Gödel's expectations from work on intermediate Turing degrees for cardinal arithmetic:
(iii) perfect set forcing (positive),
(iv) greater symmetry between sets and mappings in degree theory of r.e. sets (negative).

(b) *Tricks of the memory*, especially in relation to Gödel's work on L and Hilbert's sketch of a model for CH in *Über das Unendliche*
(i) notes in the Nachlaß of a lecture at Brown University in 1938,
(ii) Bernays' review of Gödel's announcements (9),
(iii) differences from Hilbert's sketch,
(iv) objective reasons for *not* drawing attention to similarities.

(c) *Documentation,* especially, by letters versus *impressionistic anecdotes*
(i) Generalities on political history and the history of scholarship.
(ii) Bertrand Russell on Leibniz's letters.
(iii) Personal elements in overtly scholarly publications.
(iv) An anecdote on the misinterpretation of letters.

(d) A successful *use of letters:* Gentzen's correspondence, with Bernays, on his original consistency proof [Ge]:
(i) Bernays' vague memories,
(ii) corrected by Gentzen's letters to him, now available at the ETH. Digressions on Gödel's
(iii) extension in (13) to all finite types of the use in [Ge] of functionals of lowest type (the main issue in (ii) above), and on his
(iv) attempts of assimilating bar recursion of finite types to bar recursion of lowest type; cf. 4e.

(e) *Applying the lessons of* (a)-(c) *in this Note to* reading *the present article.*
(i) Virtues of a conventional style: recognized, but not followed,
(ii) partly in view of reactions to the style of [RS].
The prominence given to problematic points inGödel's work is related to
(iii) the specific effectiveness of — most parts of — Gödel's expositions, and generalities about progress in the sciences compared to the arts and literature (with obvious consequences for the tradition of going back to the sources),
(iv) the writer's non-academic philosophy of knowledge, and
(v) his attachment to his own particular view of personal friends, concluding with a frivolous note
(vi) on the joy that Gödel's formal errors have given to others.

The occasion demands a topic that spans, if possible, Gödel's working life. For me this singles out the area of intuitionistic logic (i.l.). Though i.l. was the main subject of our frequent conversations over some 15 years, I neglected it in my otherwise detailed obituary for the Royal Society [RS]; in accordance with its motto *nullius in verba*, which I take to mean that effective contributions to scientific knowledge are to be given priority over reflections about the latter. In [RS] formal systems and sets were stressed, with some attention to various kinds of definability since they have turned out to be more effective scientific tools than the intuitionistic notions; even for purposes that the latter have been claimed to serve.

Specifically, in connection with *constructivity*, the principal element of i.l. — of applying this requirement to proofs, and not only to definitions — is highly dubious; cf. [KM], but also Gödel's footnote on p. 447 of [BP]. In fact, in the forties, Gödel emphasized definitions; both in connection with the ramified hierarchy (of the constructible sets, loc. cit.) and with logic—free, essentially equational systems discussed in section 4 (incorporating his successful experience with higher types, cf. top of p. 197 in [RS]). In contrast, in the early 30's, as developed in section 2, Gödel looked at typically intuitionistic, logical aspects of constructivity, and — as in other areas — returned to these interests in the seventies, albeit in a different style; not the early free display of sound intellectual reflexes, but reverent attachment to the tradition of academic philosophy.

It turns out that i.l. generally, and Gödel's contributions to it just mentioned can be used very effectively for a part of knowledge that is outside science, and incidentally quite close to the popular meaning of 'philosophy'; cf. [K17] for more details. Roughly speaking, as elaborated in section 1 below, here the primary object is not merely to pursue traditional ideologies, and problems fundamental to them, but to examine them. And i.l. provides so to speak chemically pure specimens of such ideologies.

GÖDEL REMEMBERED
Salzburg 10-12 July 1983
Ed. by P. Weingartner and L. Schmetterer
BIBLIOPOLIS 1987

The main text below, and the two appendices are organized accordingly. (The latter point out a neglected aspect of Gödel's style in the 30's, resp. a parallel between the high spots of classical logic 1880-1930 and i.l. 1930-1955.)

The notes at the end contain reminiscences of Gödel, starting in the mid fifties. They may balance the picture of him provided by recent Gödeliana, badly weighed down by material from the seventies, cf. top of p. 160 in [RS]. Readers familiar with this stuff will notice that some of its silliest items are debunked by immediate corollaries to observations below, with a sort of summary in (d) at the end of App. 2. Charity forbids my giving chapter and verse; anyway, in most cases I have forgotten them.

1. Background and a manifesto

Looking back at the early thirties, when Gödel did his best known work on i.l., the first order of business was to help clean up the logical pollution spread by Brouwer and his epigones; comparable to Hilbert's logical atrocities with his claims for consistency; cf. Note 1. Gödel's distant, sometimes almost offhand (*hochnäsig*) style still seems fitting; cf. bottom of p. 153 of [RS] on consistency. Time, if not Gödel's style, has dispelled the pollution, and it is appropriate to look at another side of i.l.

To put first things first, it is most memorable as a reaction to an earlier, also would-be revolutionary, but by [K14] not too different enterprise, socalled logistic foundations. Among other things they were said, for example, by Russell to have exhibited the concepts implicit in our ordinary logical reasoning. Granted, logistic, that is, truth functional logic may be better — for example, for reasoning well — than the logical notions of natural language. But they are certainly different! even in the propositional domain. Thus $(p \to q) \vee (q \to p)$ is not a law on any ordinary reading of $\to$ and $\vee$. In large domains of *natural mathematical language* $\exists$ is taken to mean that an instance can be defined, tacitly, in terms used in that domain. For better or for worse $\exists$ is not regarded as an abbreviation for $\neg\forall\neg$. Without exaggeration, i.l. wanted to approximate natural usage better, and succeeded.

So much is clear, though not often stressed. Much more interesting for the kind of philosophy adumbrated already (and explained in [K17]) are early impressions, *alias* convictions, about i.l. They were particularly *off the mark where all sides agreed*; specifically, about some intrinsic complexity of i.l., and about — what we should now call — its proof-theoretic weakness. The faithful were prepared to pay this price for the sake of the Truth, critics saw this as a reason for ignoring i.l., generally without investigating further. (And as we shall see in 2a, some of those who did investigate, did not pause to reflect on the implications.) Incidentally, history repeated here the experience with ramified types during the first decade of this century, when all sides agreed — of course, on complexity (of the unfamiliar, as in (ii) on p. 186 of [RS], but cf. also the end of Note 1) and — on the axiom of reducibility as its chief embarrassment; it turned out to be true for cardinal levels; cf. p. 198 of [RS].For an examination of ideologies, as at the outset of this article, systematic oversights are just as relevant as any successes. — One more reminder:

The implications or, as one says in mathematics, the corollaries are obtained from formal investigations by one-liners. Nevertheless experience shows that very often the investigations are not demanding, being done independently by different people, and it takes years before the implications are noted. In Hilbert's terms, here the latter are the building (Bau), and the formal work is the scaffolding (Gerüst).

Some logical implications of the principal point above where i.l. is presented as a better approximation to — the logical features of a dialect, as it were, of — natural mathematical language: i.l. as a standard of reference for studies of other parts of other natural languages. Generally, (a) a look at the sophistication and elegance of i.l. during the last half century allows one to see the *threshold* where its study began to touch essentials; both regarding its relevance and its limitations. In the light of i.l. other linguistic studies can be viewed realistically; not as alleged pioneer work, for which only quite lax standards would be appropriate. More particularly, (b) the main limitation of i.l. is not — contrary to what is often claimed — the 'internal' difficulty of a formalization or of finding adequate (intended or contrived) meanings, but the superiority — again, for intended or discovered purposes — of *paraphrases*. (This is quite consistent with i.l. being elegant and satisfying, a virtue of all successful *jeux d'esprit*.) Last, but not least, (c) i.l. provides a lively reminder of the fiasco of *natural history*, which studies phenomena that strike our untutored attention; in contrast to the now dominant scientific tradition, which relies heavily on 'artificial' constraints imposed by experiments. Here phenomena are isolated that lend themselves to rewarding study (as always, by something like available means).

(c) above is developed at some length at the end of section 2 after a closer look at some classes of propositions (and operations on them) for which i.l.,but not classical logic holds. In terms of (c), they are kinds of propositions that strike our untutored attention, and incidentally have done so since Aristotle. NB. However, the classification cuts across such familiar grammatical categories as declarative sentences etc.

Conversely as it were, general considerations on natural languages throw light — in an obvious, but neglected way — on the socalled creative subject, a central element of the intuitionistic enterprise; see (c) in section 3.

Advice to readers. It is probably best to leaf through the text below, letting the eye pause on titles in capitals, and to make selections for further reading. It is in the nature of the case that the range of interests, even within i.l., of an exceptionally gifted person like Gödel will be exceptionally broad, and so selection is advisable.

2. Early metamathematics of systems for i.l.

In the early 30's Gödel published a few notes on i.l. As is to be expected the results are superseded; incidentally, some of them were not demanding even at the time inasmuch as they were found by others independently. The style is refreshingly concise, almost offhand; in obvious contrast to the then prevalent heavy rhetoric of i.l.; cf. Note 1. Less trivially, the notes are widely accessible; they — like incidentally most of Gödel's publications — require little mathematical background. The price is high: the notes provide little indication of any domains where the result might be really relevant. — His comments, right or wrong, are almost uniformly rewarding as a reliable record of first impressions (on i.l.).

Below, most results are stated for the *propositional* part, which illustrates very well many properties of i.l.; cf. the very successful chapter 1 of Chang/Keisler for classical logic. In i.l. the propositional part becomes particularly rewarding when its *quantifiers* are included. (Propositional quantifiers bring little in the classical case, at least, generally; there are some exceptions in socalled complexity theory.)

a) The *negative fragment* consists of the operators ¬ and ∧ (with ∀ in predicate logic) applied to negated atomic formulae. By (5)* the same formulae of this fragment are derivable in — the usual systems of — classical and i.l. Further, there is a quite efficient transformation of any classical derivation into a, generally different intuitionistic one (with the same end formula).

Since the negative fragment is a socalled reduction class for the full *classical* fragment, the latter is thus embedded in i.l. (preserving most relations prominent in metamathematics). However, interpretation and scope of this kind of embedding are a delicate matter. Above all, there is the question:

What is gained by having an intuitionistic rather than only a classical proof of a negative formula?

Evidently, granted intuitionistic ideology the answer is trivial: now one has a *valid* proof. For philosophy in the sense of section 1 this answer is worse than useless. It stops one from even looking for a convincing answer, for example, in terms of functional interpretations of formal derivations. (Reminder: derivations in i.l. are usually realized by operations that are continuous in some suitable sense, and so the translation ensures that an arbitrary realization can be replaced by a continuous one; cf. first-order formulae about real closed fields, where any realization can be 'replaced' by an algebraic one.)

Gödel's early notes do not touch the question above. But they raise quite a number of less delicate points that are still of interest.(Reminder: Gödel's own answers usually reflect only first impressions; not only his, but — as shown, for example, by Skolem's review of (5) — also by distinguished contemporaries.)

(i) Concerning *differences in meaning* between classical and intuitionistic operators Gödel thought they were mild (gering abweichende Interpretation). This is like saying that the notions of countable and uncountable structures differ mildly because the same first order formulae are valid classically for both classes of structures.

If — following Menger, cf. Note 1b (iv) — one wants to dismiss i.l., one has to find a less hackneyed (metamathematical) property than

* Numbers in round brackets refer to Gödel's articles in the bibliography.

conservation of classical logic over i.l. for the negative fragment. Incidentally, later Gödel became supersensitive about differences in meaning; cf. Note 5d.

(ii) Concerning the scope of his result Gödel said at the end of his note that it might *fail for* socalled *impredicative systems.* This is doubly wrong. First, the proof extends almost verbatim to the theory of species, and with a little care in choosing — among classically equivalent — formulations, also to set theory. More subtly, as Gödel had observed himself, the result extends to formal classical number theory though the latter isn't all that predicative either, at least, in the strict sense. Specifically, an object may be defined by a quantified property A, that is, for which $\exists!xA$ is derivable, without there being a numeral $\bar{n}$ for which $A\ [x/\bar{n}]$ is derivable (where numerals are typical of definitions 'independent of the totality of all natural numbers'). This is a corollary to the incompleteness phenomena.

But much more significant is the following oversight. Gödel had swallowed the then — and incidentally still — widespread superstition, mentioned in section 1, about i.l. lacking proof-theoretic strength. Accordingly he never noticed that this superstition was refuted by his embedding! It was a bewitchment, but not primarily by (clumsy?) language. — The next item is for reference in Appendix 2, A2 for short.

(iii) *Corollary* of the embedding: *propositional logic is complete for the negative fragment* and predicate logic is weakly complete, that is, the double negation of completeness holds.

The proof needs nothing beyond the embedding except the realization that only very special kinds of propositions and predicates are relevant; cf. the digression in A2a. Yet the corollary was not observed for a quarter of a century after Gödel's note on the embedding. — This is fitting for somebody who, like young Gödel, regards the intended intuitionistic meaning as illegitimate or at least as sterile, and does not want to sanctify it by proving completeness for it.

Beyond the negative fragment

on the principle: What do they know of England who only England know? the heart of i.l., as already noted in section 1, is outside the negative fragment: even intuitionistic rhetoric is dominated by talk about $\exists$ together with v; cf. Note 2. Indeed, in connection with (ii) above, Gödel's result fails for systems with *function symbols,* which involve so to speak 'hidden' $\exists$-symbols. Reminder: various forms of the axiom of choice are quite weak when added to systems of i.l., but their negative translations are not, for example, in Spector's work [S].

The most fruitful exception concerns $\forall\exists$-*theorems,*[1] to which the conservation result for the negative fragment has been extended, in various ways, over the last 30 years. These theorems are, without exaggeration, typical of *algorithmic propositions*, prominent in intuitionistic rhetoric. Once again the interpretation of the results is more demanding than the proofs, the latter having often been found independently by several people. — There were two stages in the interpretation.

First, following the preoccupation with proof-theoretic strength, in (ii) above, there was emphasis on the 'adequacy' of i.l.; in particular, for proving that a program is totally defined; cf. section 3 on effective rules for the perfect digital computer, and the distinction between (equational) programs with number e for which $\forall x \exists y T(e, x, y)$ can be proved classically, resp. intuitionistically; here Kleene's T-predicate is meant. For systems that 'correspond' in a natural and precise sense the same programs are provably effective. This establishes adequacy of i.l. as understood in the foundational tradition; cf. Note 1c.

More recently, and in line with the manifesto of section 1, it was realized that

adequacy in the foundational sense *ensures algorithmic inadequacy.*

[1] The results are also known under the proprietory names (i) Markov's rule and (ii) Markov's principle: (i) if $\forall x (A \vee \neg A)$ and $\neg \forall x \neg A$ are derivable so is $\exists x A$, resp. (ii) $[\forall x (A \vee \neg A) \wedge \neg \forall x \neg A] \rightarrow \exists x A$. Their validity depends on the kind of predicate A considered; for example, neither holds for A with lawless parameters. — Incidentally, in accordance with the introduction to section 2, there is a neat propositional analogue, if $P \vee \neg P$ and $Q \vee \neg Q$ replace $\forall x (A \vee \neg A)$, $\neg (\neg P \wedge \neg Q)$ and $P \vee Q$ replace $\neg \forall x \neg A$, resp. $\exists x A$. The principle is not generally valid. Closure under the rule follows by (an ad hoc) use of the disjunction property or, at the other extreme, by specializing some general fancy proof of the kind discussed in the next paragraph but one.

In fact, there are relatively simple proofs, say d, of $\forall\exists$ theorems, say, $\forall x \exists y R(x, y)$ (for example, of definition by transfinite recursion $<\varepsilon_0$) that are hard to unwind. In other words, it is costly to extract or execute a program π_d such that $\forall x R(x, \pi_d x)$: this is algorithmic inadequacy. More formally, and more neatly, recent proofs of Markov's rule by Dragalin and H. Friedman show quite generally how easily *any* classical proof of a $\forall\exists$ theorem can be converted into an intuitionistic one; so i.l. is algorithmically no better than classical reasoning (for typical algorithmic problems); cf. Appendix 4 of [K16] for more details. —

Remark. Here, in accordance whith section 1, the recent efficient transformations: $d \mapsto d_i$ (of a classical proof of $\forall\exists R$ or of an intuitionistic proof of $\forall \neg \forall \neg R$ into an intuitionistic proof of $\forall\exists R$) are used for a *critique* of foundational aims. Within the foundational tradition the advantage of the recent proofs of Markov's rule over earlier proofs would be seen in the use of more elementary metamathematical methods. According to section 1 this traditional view is itself distinctly problematic as long as there are no realistic doubts about the old methods, restrictions being as good candidates for justification as extensions. — This remark brings us to a venerable

General worry: a seesaw in interpretations or shifts of emphasis, Akzentverschiebung in psycho-analytic jargon. Just now, two stages in the interpretation of closure under Markov's rule were mentioned. *Where will it all end?* After all, the first interpretation as establishing adequacy (or stability of the notion of provably total recursive function) had a comforting look of finality about it.

The problem is genuine, and perfectly well recognized under such code words as 'dialectics'. But all these dialectical fireworks draw attention away from the *facts of scientific experience.* Time and again interpretations have settled down, just as expositions of various branches of science quite often reach a stable form, to be enriched, occasionally, by a more sophisticated vocabulary. For example, already Goursat's *Cours d'Analyse* grouped the elementary parts of the subject in more or less the current order, except that today we give *names* to those groups: theorems valid in all Frechet spaces, topological spaces, metric spaces and so forth; realizing as it were the biblical idea of paradise, in Genesis 2, 19, where God brought Adam the objects He had created, and Adam gave them names; presumably, thus coding — his knowledge of — the principal properties of those objects.

Incidentally, Gödel himself had a horror of shifts of emphasis — which he would have called 'shifting one's ground', if the question had arisen —, and saw in them a principal reason why philosophy made so little progress. Without exaggeration, it is more likely that attachment to a few — sterile — interpretations and to problems fundamental for them has hampered the progress of philosophy than too many or too imaginative shifts of emphasis. (At least occasionally the silent majority's practice of ignoring those interpretations is tantamount to attachment by — benevolent — neglect.)

(b) *General provability and formal derivability.* Gödel's note (4) contains a translation of intuitionistic propositional logic into one of the systems of (classical) modal logic. The additional operator $\Box$ is variously interpreted as some kind of necessity or provability; more on this is to be found in (iii) below.

As it stands the note does not go far. Gödel had simply focused on one item in the rhetoric of i.l.; in particular, on the alleged opposition between truth and provability or, in modern jargon, between truth and assertability conditions. He then tried out the first formalism at hand with the smell of that opposition. Later the note was refined by others who showed that the translation was faithful. Gödel's own result was enough to establish simple metamathematical properties of i.l., for example, the underivability of $p \vee \neg p$ by Heyting's rules.

But the note is a good peg on which to hang various observations of more permanent interest.

(i) *Soundness and incompleteness.* Gödel himself used $\Box$ to improve his original formulation of the *first* incompleteness theorem. Instead of talk about the Liar, which sounds merely frivolous to most of us, Gödel here interprets his independent sentence as an instance of the most natural property in the world, soundness (also called reflection principle):

$$(\Box p) \to p\,,$$

where $\Box$ is now formal derivability in Principia Mathematica (and related systems). This form establishes incompleteness for intuitionistic systems of arithmetic too in the sense that some valid sentence is not formally derivable; evidently, here it is not enough that some sentence be formally independent. The formulation is also superior to the second incompleteness theorem, about consistency; first, in being applicable to a broader class of systems, for example, not requiring demonstrable completeness w.r.t. Σ_1^0 sentences, and secondly by referring directly to $\Box p \to p$ (for $p \in \Pi_1^0$); this is ensured by consistency (modulo Σ_1^0 completeness), and is the only reason for regarding consistency as sufficient for any kind of soundness.[2]

(ii) In a different vein Gödel's note remains of interest because it isolates a property, in the elementary formalism of modal propositional logic, that distinguishes between formal derivability and general provability; most elegantly, by use of Löb's theorem. Not only is some instance of $\Box p \to p$ underivable, but one has

(*) $$\Box(\Box p \to p) \to \Box p \;.$$

This is an opposite extreme as it were of the property $\Box(\Box p \to p)$ of general provability since the latter, together with (*), implies $\Box p$.

(iii) *Truth and general provability;* at least so far, a distinction without much difference. Formally, all axioms of the modal logic considered by

[2] The points just made deserve comment. (a) At this point, completeness is not required for all Σ_1^0 sentences, but only those expressing formal derivability. As A. Visser has pointed out to me, contrary to (my) first impressions, not all Σ_1^0 sentences are demonstrably equivalent to some $\Box A$. In fact, suppose, for some R (not necessarily in Σ_1^0), $\Box R \to \Box \bot$ *is derivable. Then,* $(\Box A) \to R$ *is not derivable.* If it were, $\Box\Box A \to \Box R$, and so, by the assumption on R, also $\Box\Box A \to \Box\bot$ would be derivable. Since, generally, $\Box\Box\bot \to \Box\Box A$, $\Box\Box\bot \to \Box\bot$ would follow. The letter 'R' is chosen because — the Σ_1^0version of — Rosser's sentence satisfies $\Box[(\Box R) \to (\Box\bot)]$; but cf. (iv) below. (b) The second point, about consistency, is implicit in one of Gödel's notes to (13), from the seventies, under the embarrassing heading 'The best and most general version of the unprovability of consistency in the same system;' 'embarrassing' because (i) it refers to *formal* systems, and so is obviously not most general, cf., for example, Mostowski's monograph [M], and (ii) it gives no hint under which conditions this version, that is, $\Box p \to p$ for $p \in \Pi_1^0$, is equivalent to consistency; so, far from being best it is not even good. — The best that could be said is that it is the version most directly relevant to Hilbert's program where Π_1^0 sentences are privileged. So it should be noted that, for formal derivability $\Box$, $(\Box p \to p) \in \Pi_1^0$ if $p \in \Pi_1^0$.

Gödel, in other words, all properties of general provability formulated there, remain valid if '□' is dropped altogether. Admittedly '□' cannot generally be introduced, for example, in the conclusion of $p \rightarrow p$, and preserve validity. But nothing is explicitly formulated about general provability that does not also hold for truth; in contrast, for example, to (*) in (ii) above for formal derivability □, or to a language with propositional quantifiers since evidently $\neg \forall p\ (p \rightarrow \Box p)$ holds, but not $\neg \forall p\ (p \rightarrow p)$. In fact, it seems to be open whether anything can be said about general provability in the language considered that does not hold for truth. Of course this will not be (mis)interpreted as showing that the two notions have the same meaning! cf. 2a (i).

It was a genuine discovery of the 70's to recognize that formal derivability admits a neat theory at all. This was *philosophical* progress, correcting the simple-minded view that general provability 'ought' to be studied. — Incidentally, the kind of general provability meant here is not likely to be concerned with the outer limits of provability. The latter grow, and so it would be *prima facie* inappropriate to apply classical logic to statements containing □ (which is not to say that therefore i.l. is appropriate!!).

The selection of rewarding phenomena among those that present themselves to our untutored attention — here, of formal derivability within general provability — is a recurrent theme of this article.

(iv) A blind spot and some technical remarks. The former concerns the thoughtless — and by now largely forgotten — literature against *mixing object language and metalanguage;* as if the union of two sets were not a set; cf. Gödel's own sensible examples in (11), about every sentence containing a relational word, and the like. But none is as memorable as the modal language considered above, particularly, when □ is interpreted as formal derivability. — Naturally, as with other unions of two sets, for example, of cabbages and kings, there is a genuine problem of finding non-trivial laws that hold for the union (in the particular language considered).

The technical remarks concern the elegant theories of formal derivability. It is an undoubtedly memorable fact that the 3 axioms expressing the (distinctive) theorem of Löb, closure under modus ponens, and completeness, at least for those Σ^0_1 sentences that express formal derivability should axiomatize all valid theorems of the language; and uniformly for a broad class of — formal and some other — systems at that. But at least so far this axiomatization has not helped to find new memorable theorems in the language itself comparable, for example, to earlier observations about the negation of consistency being conservative for Π_1^0 sentences (which can be done in the language for those expressing underivability; cf. (a) in footnote 2). One would have hoped that by completeness a property established for some cunningly chosen system could then be generalized to all formal systems considered; as properties of the field of real numbers are generalized to all real closed fields. The second technical remark is a reminder. Though cut-free systems have been recognized to be significant for current logic, and by [KT1] to have memorable properties in the language considered, little is known about formal theories for cut-free derivability. — A titbit: without recognizing the general significance of cut-free derivations, von Neumann is on record as having seen the principal difference between his and Gödel's proofs of the second theorem in these terms.

(c) *Infinitely many monadic propositional operators.* Gödel's halfforgotten note (6) states that Heyting's calculus for $(\neg, \wedge, \vee, \rightarrow)$ does not have an adequate finite truth table. Now, infinite truth tables, such as boolean algebras, are just as good or better, for example, for decision procedures, the most prominent purpose of truth tables 50 years ago. Here there is no particular virtue in a finite truth table for the *whole* language; success requires an efficient way of first finding one for any given formula, and then evaluating it. So the result stated in (6) is obsolete.

Of more lasting interest is a step in the proof that provides a sequence, say A_n, of pairwise (formally) inequivalent formulae with just one propositional variable p. In other words, there are infinitely many monadic operators: $p \mapsto A_n$. Or, more pedantically, they are different for all classes of propositions and interpretations of the operators $(\neg, \wedge, \vee, \rightarrow)$ for which the calculus is complete. — As a memorable corollary there is a sharp contrast with the classical case where there are just 4 monadic operators: the 2 constants $\top$ and $\bot$, p and $\neg p$.

Viewed as above, Gödel's proof suggests immediately the operator o:

$$p \mapsto WA_n \text{ for all } A_n \text{ not equivalent to } \top \,;$$

cf. [G] on the wide spectrum of meanings for which o is not equivalent to any operator built up (finitely) from $\neg, \wedge, \vee, \rightarrow$.

The remainder of this subsection, that is 2c, attempts to give some perspective on new propositional operators; in line with the manifesto in section 1, not only as a topic of logical research, but for examining ideologies. Naturally, we begin with what is known.

(i) By relatively recent work of P. Wojtylak [W1], the fragment ($\neg, \wedge, \vee, \rightarrow$, o) has a respectable metamathematical theory; cf. [W] also for references to other work on monadic operators of i.l., defined by infinitary conjunctions and disjunctions or by propositional quantification. As far as mere legitimacy is concerned, ($\neg, \wedge, \vee, \rightarrow$) is seen to be just one fragment of — the propositional part of — i.l. among many others.

The result of de Jongh [J] on the unbounded totality of binary operators is naturally interpreted by a metaphor from set thory. While a fragment is a subject for research — to be compared to a set which can be grasped as a unity — its complement, that is, the totality of new operators, is not.

For a proper perspective, here is a reminder of experience in classical logic.

(ii) The functional completeness of its propositional part is a quite exceptional phenomenon in classical logic. Thus, already when applied to sets and relations the question arises which properties distinguish the boolean operations, say $\cap$ and C among all operations on the collection of subsets of a given set V, resp. of some Cartesian power of V; cf. Craig's attempt in [Cr.]. — The obvious parallel in i.l. comes up in the socalled topological interpretations with operations on *open* sets.

The following points, concerning classical predicate logic, are in keeping with the familiar fact that the propositional part of i.l. exhibits many of the formal complexities of classical predicate logic. So the reference below to socalled abstract model theory is not dragged in out of the blue. Incidentally, another fact of experience — how quickly abstract model theory rose and fell in the 70's — is in keeping with (i) above about it not being a subject at all.

The best known result here is Lindstrom's maximality property of the classical fragment ($\neg, \wedge, \forall$), to be compared to attempts of sanctifying the fragment (¬, ∧, ∨, →) in i.l. The comedy involved has been described often enough; perhaps, most recently in the review of [SV]. It need not be repeated here except perhaps for this. Relevant extensions of experience such as considering new quantifiers or fragments of $L_{\omega_1\omega}$ (and remembering those extensions!) have been more rewarding than the kind of brooding common in the socalled theory of meaning.[3]

Last but not least there is a broad parallel suggested by the (general) view of i.l., for example, in [K14], as being concerned with a class of propositions beyond those of classical logic; following Aristotle, cf. Met Γ 5, 1009a, 16-22 or Met Γ 7, 1012a, 21-23. Incidentally, a compact formal expression of this view is found in old-fashioned systems of i.l., with modus ponens as its only rule of inference and literally a subset of the axioms for classical logic. This then evidently allows for more interpretations with larger ranges for the (propositional) variables.

(iii) Propositions and numbers: some parallels, especially for orientation on new propositional operators. Leaving aside pretentious drivel about the origin of 'the' concept of number — at least, till somebody has as imaginative an idea, mutatis mutandis, as Darwin — one may think quite reasonably of various kinds of numbers that populated the intellectual life of the 18th or 19th century; including mildly embarrassing names like 'real' and 'imaginary', resurrected in Hilbert's terminology of real and ideal elements. The traditional perennials about existence, subsistence or what have you of those numbers simply draw attention away from the work that has been done about them. Nothing comparably imaginative has so far been done with propositions. So to convey the parallel in question it is best to begin with some reminders about numbers; specifically, about successful choices of particular kinds of numbers.

Most familiar is the series of socalled extensions (of the number system) by means of closure properties required for solving larger classes of problems; for example, from natural to whole to rational to algebraic to real and complex numbers, and so forth. 'Socalled' because, if numbers are thought of in connection with length (as they were in Euclid, not counting),

[3] For examples of such sterile brooding readers may look at [Gol] on the meaning of the quantifier in the twenties, and, in case of i.l., at the papers of Sundholm [Su] and Weinstein [We], and their reviews.

something like the real numbers comes very early; to be *paraphrased* later in terms of suitable sets or sequences of rational numbers. So to speak in the opposite direction there was work on limits to such 'extensions'; for example, in terms of division algebras with the magic numbers (for the dimensions): 1, 2, 4, 8. Success depends on a proper selection of the properties — here, of + and × — to be preserved in the extension. Before bandying about words like 'fundamental', it is a salutary exercise to remember situations where it is relevant to think of commutative division algebras (over the reals) with a $\sqrt{-1}$, and where the geometric representation (x,y) with the familiar laws for + and × provides effective knowledge.

Trivially, there are many more kinds of numbers than can possibly be used for effective knowledge. Selection requires thought on what we need to 'do' with them. Certainly, one thing we 'do' with them is to operate on them; in short, the choice of new operations, that is, functions, is an integral part of the extension. Thus the passage from the algebraic to all real numbers would simply not be exploited well if all operations were still required to be algebraic (even in the weak sense of having algebraic values at algebraic arguments). This would exclude the exponential and trigonometric functions, since, for example, 0 is the only algebraic α for which $\sin \alpha$ is algebraic. In short, we should have

> more numbers without doing anything new with them.

In the parallel meant above, the collection of logical operations ($\neg$, $\wedge$, $\vee$, $\rightarrow$) corresponds to some familiar collection of, say, rational or algebraic functions (or literally to number-theoretic functions mod 2: $1\text{-}x$, xy etc.).

For the view of i.l. here considered, as concerning larger classes of propositions, for example, about choice sequences the parallel has an evident implication:

> If we cannot think of anything to do with new operators the chances are that *either* there is not much of interest to be done with the extended class of propositions *or* we have not even begun to understand the possibilities.

Informed readers will remember here that till the 30's most logicians had not even begun to understand the possibilities of i.l. beyond finitist mathematics. For philosophy in the sense of section 1 this fact is relevant, but also the obvious attraction of sanctifying the familiar fragment (parallel to the use of Lindstrom's theorem for classical predicate logic discussed in 2c (ii) above); cf. p. 1296 and p. 1298 [SH] concerning 'completeness', resp. 'strength' of the fragment. But the metamathematical properties involved in those notions are so hackneyed that they do not constitute any *test* of — the relevance of — the fragment; rather an expression of attachment; cf. the end of 2a on such matters.

Remark on another view of i.l. but still in the light of the parallel with numbers: proof analysis by abstraction. Here one does not think of a literal generalization, in particular, of a larger domain of objects, but views an axiomatic analysis as identifying abstract properties that are relevant to given theorems (about a specific domain); with additional information as pay-off for eliminating some axioms. For example, many elementary results about the rationals use only the field properties of Q (not even that Q is a number field). Then any Σ_1-theorem $\exists xA$ allows a sharpening to

$WA\,[x/t_i] : 1 \leqslant i \leqslant N$ for t_i depending rationally on the parameters

of A. If the proof of $\exists xA$ uses only i.l., a further sharpening is possible. A single t will do ($N = 1$); recall other existential properties from 2a and Note 2.

But after nearly 30 years of experience with this kind of search for additional information, also by others using sheaf-theoretic models (a fancy way of talking about the continuous dependence of y on x in combinations like $\forall x \exists y$), I am sceptical. At least so far one has fallen between two stools. On the one hand, when this sort of additional information is really needed i.l. is not refined enough; recall 2a on its algorithmic inadequacy. (And to come from the sublime to the ridiculous, there is nothing in the ordinary mathematical tradition to stop one from recording such information if one has it.) On the other hand the ritual of i.l. prevents one from *testing* its ideology, which, as emphasized at the outset, requires not only explicit definitions, but a constructive proof that they do their job.

So — by comparison, and for the time being — the view of i.l. as dealing with a larger class of propositions seems more rewarding; at least, for the following object lessons in section 1.

(d) *Natural languages* with some reminders on natural history. By section 1, i.l. is a successful study of — the logical features of — a popular dialect of natural mathematical language.

(i) Reference to natural mathematical language — for assessing methods of studying other linguistic phenomena — is at any rate plausible; provided of course it is remembered that such things as the language of texts on axiomatic set theory in the fifties are artifacts. Perhaps the single most significant point here is *survival value* since, by experience, this has been a successful guide in biological sudies (and natural languages are certainly a biological phenomenon). Judged by survival value mathematical language, at least of elementary mathematics, is more convincing than talk about cats doing something or other on mats; cf. W.H. Auden's reservations on the grammarians' preoccupation with such talk in his opening address to the Salzburg Festivals.

(ii) As to the success of i.l. it applies not only to the syntactic aspects adumbrated in section 1, but also to various meanings — associated with them or, more precisely, appropriate — in various situations. As in (i) a couple of provisoes have to be remembered.

At the present stage a realistic measure of success is to find relatively few meanings that are appropriate in relatively many situations; in contrast to talk about a necessarily amorphous family of meanings. Some of these meanings can be thought of as corresponding to other external 'parameters' besides the words used; such as the tone of voice, expression of face, gestures. — All this does not exclude the possibility of much more sharply defined specifications in completely different terms; comparable to those in molecular biology of family likeness such as the Bourbons' nose or Habsburgs' hare lip so central to Schrödinger's *What is Life.*

But readers should beware of the — most simple minded, and — superficial 'alternative', to the small arsenals of meanings above, of introducing an additional variable for situations (or, for that matter, tone of voice and the other external parameters mentioned earlier). This remains empty unless something substantial can be said about the situations that arise; cf. 2b (ii) on general provability or the socalled abstract theory of constructions which introduces a variable for proofs but, by p. 80 of [K14], nothing less banal about them than the relation between a proof and the assertion proved. This is as sterile as the business of situations 20 years later.

The weakness of those general schemes is underlined by the occasional particular problems that do benefit from attention to proofs — for example, *q* - and fp realizability —, resp. to situations — in which, for example, a counterfactual conditional is satisfied —. This is so because the schemes do not help spot those problems, but may distract from the fact that the latter are the exception rather than the rule.

Digression on a widespread oversight, about the skills needed to use (any) theory; illustrated, perhaps, exceptionally well in linguistics generally, and by the business of situations in particular. No phenomenon that presents itself *naturaliter* —in contrast to experiments, which are set up to exclude forces not treated in the theory at issue — comes with a label telling us *which* theory (if any) applies or, equivalently, which forces dominate it. So clearly some skill beyond knowledge of the theory itself is required; even in the case of planetary astronomy Tycho Brahe had to shift from the observed, also called 'apparent', motion to its 'correction' (for parallax), since only the latter lends itself to theoretical analysis. It may be common to assume that linguistic phenomena (and others of the socalled human sciences) are very different in this respect because we have conscious beings speak to us, and not dumb planets. But if common, it is simply a common piece of scientific immaturity. Viewed in these terms the business of situations is a *step back* from the small arsenal of meanings at the beginning of 2d (ii) above. The latter reduce the additional skills to a proper choice from that arsenal, while mere mention of 'situations' says nothing about their particular aspects that may be relevant.

The next two observations are less demanding.

(iii) It is an *illusion* — surely, not wilful deception — to present studies on natural languages as pioneer work, to which correspondingly lax standards are to be applied. This simply overlooks the fact that, realistically speaking, such subjects as i.l. are also studies of natural languages, and their level of sophistication provides more appropriate standards.

(iv) More specifically, experience in i.l. underlines the need for *selecting rewarding aspects* of linguistic phenomena. For example, such classes of propositions as considered in i.l. certainly turn up in natural languages, and many more besides. But, by 2c (iii), already those classes are suspect, as being too diffuse for rewarding theory.

This point is elaborated in (vi) below in the broader context of natural history superseded by the scientific tradition, after a brief digression for general perspective.

(v) *Natural and logical sense:* a neglected distinction. The most obvious instance is that of closure under the usual logical operations. If the propositions p and q have logical sense so has $p \vee q$; in classical logic in terms of truth values, in i.l. in terms of proof conditions. But, as a matter of simple experience, this is not so for natural sense; for example if

p = this glass is 5 cm high *and* q = this glass is transparent.

Trivially, combinations of propositions that have logical but not natural sense can be *given* logical sense (even uniquely). The question is: at what price?

As in 2c (iii) experience in mathematics seems relevant; only now propositions are not compared to numbers,but to sets (of points); as in the topological interpretation of 2c (ii). Logical sense concerns brutal existence; natural sense, for sets of *points,* involves geometry. It is a common place that not all sets of points are geometrically significant, cf. p. 213 of [RS] for references, also to Gödel himself. Readers may try out other parallels, for example, between logical sense and measurability in the sense of Riemann or Lebesgue. Incidentally, geometric sense is then not altogether irrelevant to logical sense! If sets topologically equivalent to a disc are regarded as geometrically significant, more can said about their *measure-theoretic* properties than about the class of all measurable sets.

Though the word 'natural sense' is not used in the literature I know, the idea is clearly implicit in a good deal of work on *partial* predicates and functions. — Gödel himself touches questions of sense towards the end of the introduction to the monograph (10) when explaining the relation between abstract set theory and the cumulative theory of types. If an *atomic* formula, that is, a formula $a \in b$, has no type-theoretic sense, it is declared to be 'abstractly' false, and compound formulae are then evaluated in the usual way. What he does not touch is where the convention goes wrong; this happens if p has no sense, t is the truth set, and so neither $\ulcorner p \urcorner \in t$ nor $\ulcorner \neg p \urcorner \in t$ has natural sense, but nevertheless the 'adequacy' condition of Tarski $(\ulcorner \neg p \urcorner \in t) \Leftrightarrow \neg(\ulcorner p \urcorner \in t)$ is imposed. Evidently, if such simple and familiar points are overlooked in the manufacture of paradoxes there is good reason to doubt Gödel's high expectations from a solution of the paradoxes; cf. section 8 (ii) of [W].

(vi) *Natural history* is barely mentioned nowadays, except by historians of science. Yet it presents a style of thought — or, as one says, an ideal of understanding, cf. Note 3 — which was once dominant, and still has great appeal. It relies on regularities in nature that strike our untutored attention, most often in the visual sphere. The world we see is determined by forms and colours; we recognize things in this way. For centuries, zoology, botany, but also mineralogy consisted of painstaking descriptions, and later classifications; always in these terms. It is a fact that forms and colours are particularly unrewarding or at least demanding subjects of theoretical study. For example, the relation between — the chemical composition of — a thing and its colour involves quantum theory. Data that strike us less or not at all, for example, mass and its centre of gravity or electric charge (not to speak of atomic structure) are more amenable theoretically or, as one says, are physically more important. Obviously, one can use a metaphor like Plato's cave for almost anything; but it is not too farfetched to see it as a reminder of the need for looking at informative data, and not just those (shadows) that happen to dance before one's eyes. As for appeal, it is just wonderful if, for a particular question, those things in front of us are enough; cf. also the simple-minded cult of the *black box*, and the more inspired view of the world advocated by R. Thom, and touched in Note 3c.

It is not claimed here that there are no areas of experience where the tradition of natural history is effective. What is disturbing — at least to me — is the scientific innocence of the linguistic fraternity. By and large they simply do not have a clue about the relation between their style and a well known, if obsolete tradition. A similar kind of innocence is behind the Faith in using mathematical methods. It is not this literary form which distinguishes natural history from the natural sciences, but the *selection* of phenomena treated. After all, there are some pretty mathematical formulae in natural history, from d'Arcy Thompson to René Thom. On the other hand early chemistry, in its search for chemically pure substances (and eventually culminating in the atomic view of matter), used very little mathematics.

Last but not least, natural history has an up-hill fight; it competes with our (immense) ordinary knowledge of just those aspects of the phenomena that it considers; for example, in the case of linguistics, with the works of literate people who have a genuine feeling for language. — Concerning possible uses for computer languages, 2a on algorithmic inadequacy provides an obvious warning. More generally, failure on 2 counts is to be expected: bad science because human and digital data processing ('hidden' in black boxes) are different, and bad technology because details of the hardware are not exploited efficiently. While 10 years ago such references to hardware were dismissed as red herrings, to day it is (almost) universally recognized that a good use of many processors working in parallel requires new programs. Parodying 2d (ii) about 'situations', one could introduce a variable for 'hardware' instead of looking for a small arsenal of programming languages which is suitable for relatively many varieties of hardware (incorporating subroutines) and for relatively many computational problems.

3. Effective rules

For the present it is enough to consider rules that define functions whose arguments and values are natural numbers — or even only numerals 0, s 0, ... — or (other) words over a finite alphabet such as formulae or derivations. The case of socalled higher types is reserved for the next section.

The topic of effective rules occupied Gödel throughout his life; with increased sophistication, at least, in his formulations (except of course for the lapses in the 70's, noted already on p. 160 of [RS]). Thus,to judge by footnote 18 on p. 356 of Church's [C], in the mid thirties Gödel was simply ill at ease with loose talk about effectiveness, while 30 years later he was ready to make explicit distinctions; cf. Note 4. Today we can be more explicit still.

Effectiveness involves reference to the systems for which a rule is meant; or, perhaps more correctly, to our idea(lization)s of them; as always, preferably with a few kinds of such systems being adequate to many situations. A by now familiar, particularly elementary kind is — our idea of — a socalled real time digital computer. Gödel's own interests lay elsewhere.

They will be examined below under 3 headings: (a) equational rules, (b) computation in formal systems, and finally (c) rules effective for the perfect mathematician. A particular subspecies of (c) is perfect in intuitionistic eyes, and called'creative subject' in the literature. It should be noted straightaway that i.l. enters in a quite trivial way into (a) and (b), via the difference between classical and intuitionistic proofs of the $\forall\exists$ theorems expressing the termination of formal computation procedures; cf. 2b and Note 2 on Markov's rule. But (c) is absolutely pivotal for anything remotely like the original intuitionistic enterprise; not, as is sometimes thought, a marginal aberration of the aging Brouwer. As a corollary, any reservations about the business of the creative subject put *ipso facto* anything like the intended enterprise in question.

(a) *Equational rules* define a, say, monadic function f by use of auxiliary functions $f_1, \dots, f_p$ (the constant 0, and the successor s above) in the form of a finite system of equations $E(f, \vec{f}, \vec{x})$, where $\vec{f}$ and $\vec{x}$ are $(f_1, \dots, f_p)$, resp. a sequence of numerical variables. One familiar sense of E 'determining' a function f requires only that $\exists! f \exists \vec{f} \forall x E(f, \vec{f}, \vec{x})$. This is generally not enough to compute a value, say $\bar{m}$, of $f(\bar{n})$ from finitely many substitution instances of $E(f, \vec{f}, \vec{x})$, that is, from a conjunction

$$(*) \qquad \bigwedge E(f, \vec{f}, \vec{\bar{x}}_i) \qquad 1 \leqslant i \leqslant N_n$$

for a suitable N_n; here $\bar{n}$ denotes the numeral with value n, and $\vec{\bar{x}}_i$, appropriate sequences of numerals (in place of $\vec{x}$). For example, if E is $f(x) = 2f(sx)$, only the constant $f\colon x \mapsto 0$ satisfies $\forall x E$, where $x \in \omega$ and $f\colon \omega \mapsto \omega$. But each finite set of substitution instances, say $0 \leqslant x \leqslant N$ is satisfied by any $f\colon f(x) = 2^{M-x}$ for $M \geqslant N$ (and by the totally undefined function to boot).

In symbols, the sharpened requirement on an *effective* system E is

$$\forall n \exists \mathscr{C} \exists! m \forall f \forall \vec{f} \; \{ \bigwedge E(f, \vec{f}, \vec{\bar{x}}_i) \rightarrow f(n) = m \} \qquad 1 \leqslant i \leqslant N_n$$

where $\mathscr{C}$ ranges over (finite) sequences $(\vec{\bar{x}}_1, \dots, \vec{\bar{x}}_{N_n})$.

Gödel appealed to a more or less arbitrary calculus to derive $f(\bar{n}) = \bar{m}$ from (*) or, equivalently, from $E(f, \vec{f}, \vec{x})$ with free variables $\vec{x}$ (and, of course, appropriate $\mathscr{C}$). For the tradition of socalled informal rigour it is more satisfactory to note that, for given $\bar{n}, \bar{m}, \vec{\bar{x}}_1, \dots, \vec{\bar{x}}_N$

$$(**) \qquad \forall f \forall \vec{f} \rightarrow [(*) \rightarrow f(\bar{n}) = \bar{m}]$$

is relatively easily decidable, to construct accordingly an equation calculus that is demonstrably *complete* for (**), and work with it; cf. [KT], with refinements in [Ro] and p. 908 of [B] concerning the use of monadic $\vec{f}$, resp. quantitative properties of the calculus. – *Digression* on a kind of opposite extreme, avoiding the artifact of choosing any privileged calculus. The notion of *effective* system describes the class of functions computable from equations without reference to any computation rules. (Instead of 'computable', 'finitely determined' recommends itself; cf. 'validity' in logic in place of 'provability'.) We now return to the matter at hand.

Since the requirement on effective E is Π_2^0 in (**) and (**) is decidable, the f defined by such E are recursive (pedantically, for some other usual definition of 'recursive'). The converse is read off from Kleene's normal form.

There are also refinements, of this kind of equivalence, in the literature on — what has been called — Church's superthesis, cf. [7]. Gödel himself avoided such matters; from his point of view wisely. Once one begins to look at particular sets of rules one inevitably sees how little one knows of the totality of possible rules or even of what one wants to know about them; no matter how nicely the particular sets considered behave; cf. also 3b (i) *in fine*.

(b) *Formal computability* by use of socalled entscheidungsdefinite or, more simply, invariant expressions; originally, w.r.t. PM and related systems; cf. also supplement II of Hilbert-Bernays, which, according to Bernays, incorporates several suggestions made by Gödel during their transatlantic crossing in 1935. — Reminder for — characteristic functions of — predicates:

The expressions considered are formulae F with a single free variable such that, for each $n \in \omega$, *either* $F(\bar{n})$ or $\neg F(\bar{n})$ is derivable, with an obvious variant for functions: $\omega \mapsto \omega$.

Two novelties, compared to (a), should be noted.

(i) Before the 30's, formal rules of inference were thought of primarily as means for *checking* rather than generating derivations, let alone, computations. Understandably, since the procedure involved in formal computation above is quite unrealistic. All derivations are thought of as laid out in ω-order, and the computation consists in looking for the first derivation whose end formula is $F(\bar{n})$ or $\neg F(n)$.

(ii) Compared to the specific equation calculus in (a), the notion of formal computation has a glamorously general look; even if one considers only (consistent) finite, but otherwise arbitrary extensions of some given formal system like PM. Gödel's afterthought in (7) on *absoluteness* should be viewed in this light. The general idea, as indeed the idea of speed-up by use of new 'abstract' axioms (in the logical sense of involving higher types), is an *Aha-Erlebnis* for all of us. But Gödel's early formulation is simply clumsy.

For one thing, the speed-up is illustrated most simply by any undecided formula $(\forall x \in \omega)\,[f(x) = 0]$. Computation of f according to its defining equation is slow. If, by use of new axioms, we know $(\forall x \in \omega)\,[f(x) = 0]$ we have the unsurpassably fast computation: $x \mapsto 0$. The more elaborate formal exercises of (7) fall between two stools; they are superfluous for the general point, and they do not help to discover realistic possibilities of speed-up. (Of course, the exercises are more 'weighty' than the aside above.)

As for the absoluteness property it is most impressive by comparison with the embarrassing drivel about the 'evidence' for Church's thesis allegedly provided by the equivalence between different definitions of effectiveness; without a thought of any safeguard against a systematic oversight. (If each juror has a chance 1/2 of judging correctly, why are verdicts wrong more often than once in 2^{12} cases?) In contrast, absoluteness constitutes a striking *closure property* of the class of formally computable functions, actually, not only for finite extensions, but all those enumerated by functions that are invariantly defined in a system already recognized as formal. — This 'raw' attraction of absoluteness goes well with the use made of it later by Tarski to transfer recursive undecidability results to large classes of systems; cf. the monograph [TMR] by him, Mostowski and R.M. Robinson.

On the other hand Gödel let his enthusiasm run away with him when he claimed in (7), and especially (14), that formal computability was unique among epistemologically interesting notions by being absolute in the sense of being independent of the language considered. What else is the word *functional completeness,* as applied, for example, to classical propositional logic, about? cf. 2c for the contrast in the case of i.l. Of course, in conversation Gödel agreed that he had had a blind spot. But he is not alone in having forgotten the great impression — at least, on logically sensitive people — when we first learn such an easy and convincing answer to the question: What is a propositional operator? but cf. Note 3a (iv) on worries about this answer being so easy that it is liable to be singular.

Remark for readers interested in philosophy in the sense of section 1. The drivel above about 'evidence' for Church's thesis obscures a genuine virtue of having many equivalent definitions or, more simply, descriptions of the same notion (whether or not they define the originally intended matter). When solving problems about the notion, use can be made of knowledge of the different concepts involved in those descriptions; cf. also c(iii) of Appendix 2. It is an object of research to *discover* which descriptions suit particular problems, even though it may well be that other descriptions tend to force themselves on us. Intensional logic, which is preoccupied with those other descriptions, is thus not an illusion, but often simply sterile.

(c) *The perfect mathematician* is usually presented as an immensely subtle idea, and rules effective for that animal — but, tacitly, not for digital computers — are sought in the outermost reaches of Higher Thought. In fact, practically none of the rules used every day, and thus stated in some natural language is literally effective for any digital computer. The discovery that many can be replaced by different rules so as still to define the same function — generally, without preserving the computation processes even approximately — was the sensation of formalization 100 years ago, and is an essential ingredient of the computer business. Programmers are paid, sometimes handsomely, to find suitable replacements.

The idea that those rules in natural language including natural mathematical language are defective for brutal reasons such as lack of precision or some other unreliability is sheer dogma. Of course, they are not formally precise. But how adequate is this idea(lization) of precision for a realistic view of reliability? cf. Note 4c (i) for a striking *petitio principii* in this connection.

The problem is elsewhere, and readers of 2 (d), but also of 2 (b) on general provability, are prepared for it.

What of interest can be said about the perfect mathematician?

and the choice of concepts or 'language' in which this information is to be expressed is part of the problem.

Here is some background, necessarily less banal than the business of cats on mats in 2d (i), including some of Gödel's own ideas. Reminiscences of conversation with him on and around the topic are in Note 4.

(i) For the intuitionistic idea of the perfect mathematician, the socalled creative subject, the following type of non-mechanical rule has become standard since the end of the 60's. It maps formal derivations d of existential formulae $\exists xA$, possibly with parameters, built up according to some intuitionistically interpreted (possibly formal) systems into terms t such that

> t defines the object or family of objects x (satisfying A)supplied by the proof $\bar{d}$ represented by d.

Certainly, no digital computer accepts this rule as it stands, since it requires an understanding of the map: $d \mapsto \bar{d}$, and of: $\bar{d} \mapsto x$, supplied by the interpretation of the system considered. (Digital computers do not handle *this* kind of understanding or interpretation.) So already the question *whether or not* the rule is equivalent to some computer program: $d \mapsto t$ goes beyond the domain of digital computing. 25 years ago it was conjectured that some such rule might define a non-recursive function.

There is an obvious formal parallel to the question above. It involves the arsenal of functional and realizability interpretations of various systems of i.l., but also transformations in the style of cut elimination. Each such operation 0 supplies a map μ_0: $d \mapsto t$ though, of course, not generally a definition of the object supplied by $\bar{d}$ itself; for example, if t is simply the smallest numeral $\bar{n}$, for which $A\ [x/\bar{n}]$ is derivable in the system considered; cf. 2b on the E-property.

Here it was conjectured (25 years ago) that most of the μ_0 are (even) extensionally different because the various interpretations are well known to satisfy laws not generally valid for i.l.; for example, the interpretation discussed in the next section satisfies Markov's rule, recursive realizability satisfies Church's thesis, and so forth.

However, those conjectures were refuted during the 70's. The operations μ_0 were shown to be equivalent even up to conversion, mainly by Mints. Furthermore the map: $d \mapsto \bar{d}$ could be examined by use of socalled theories of abstract constructions (of little use for anything else) with the result that the μ_0 were seen to be equivalent to the rule stated above in terms of the (intended) intuitionistic interpretation. This is the kind of safeguard against the possibility of a systematic error that is lacking in the socalled evidence for Church's thesis considered in 3b.

Perhaps the single most memorable corollary to all this *prima facie* satisfactory work on the stability as it were of the idea of the perfect mathematician is this. Once one looks closely at the map: $d \mapsto t$, one sees how marginal the algorithmic aspects of proofs are; mathematically quite trivial changes in d lead to algorithmically wild changes in t, cf. section 5 of [K16]. This fact is clearly embarrassing for several variants of the intuitionistic ideology. From *their* point of view it supports Gödel's worry, reported at the end of 3a, about leaving well enough alone; but not for philosophy in the sense of section 1.

(ii) *Wenn schon, denn schon* (you might as well be hanged for a sheep as for a lamb). Evidently, (i) imposes what appear to be gratuitous restrictions on the perfect mathematician by requiring perfection in intuitionistic eyes. In notes to the English translation of (13), from the seventies, but along lines discussed earlier, Gödel goes to the opposite extreme and considers rules of the form:

> compute the characteristic function of $\vdash_2 A_n$ for a suitable sequence of formulae A_n where $\vdash_2$ means second order validity, and $\vdash_2 A_n$ is to be decided by suitable axioms of infinity.

Gödel assumes familiarity with the subject, relying — in effect, though not in so many words — that St. Thomas' *adaequatio et rei et intellectu* would furnish the required axioms. In those same notes Gödel enlarges on that *adaequatio* by pointing out that the intellect grows (when familiarizing itself with the material at issue). However, this does not add much; *not that it grows, but how* is the crux.

In terms of Note 3b (i) it is doubtful whether our knowledge of the possibilities of the mathematical imagination has reached the threshold for pursuing any idea(lization) of the perfect mathematician. Or, to put it in current jargon, there is nothing sufficiently specific that we know about biological data processing — nothing like Planck's discovery about black body radiation in another domain — to have confidence in such ideas. The same applies, of course, to earlier jeux d'esprit in this area, now known as autonomous progressions.

4. Effective rules of finite type

This matter was the principal topic of my conversations with Gödel, which is reflected in the style of the present section.

(a) *Background.* In the first 20 minutes of our first meeting, in October 1955, he sketched some formal work he had done in the forties, and later incorporated in the socalled Dialectica interpretation (with a total shift of emphasis). He was familiar with my own interest, also since the forties, in what I called functional interpretations. They rely on a kind of $\exists\forall$ normal form where — in contrast to Skolem's normal forms — the quantifiers need range only over recursive objects, albeit of higher type.

Gödel's interest in the forties, as described to me — but also in his notes for a lecture at Yale on the occasion of his honorary doctorate —, was quite different. He wanted to fill the superficially principal gap left by his negative translation, treated in 2a under the heading: beyond the negative translation. (In his own words in his notes in the Nachlaß he wanted to find out to what extent i.l. was really constructive.) He dropped the project after he learnt of recursive realizability that Kleene found soon afterwards; cf. Note 2b (ii).

Today the relations between the two schemes are summarized by the general facts about the *E*-property in 2a and Note 2, extended in (i)-(v) below, which, for convenience, repeat some of the general material.

(i) As originally presented neither scheme achieves quite what Gödel intended. For any derivation, say d, of $\exists xA$ in the system HA of arithmetic considered, both obtain terms, say t_d^G and t_d^K (where Kleene's t_d^K is simply the number of a partial recursive function depending on the parameters of $\exists xA$) such that

> $A\,[x/t_d^G]$ and $A\,[x/t_d^K]$ hold for the Dialectica, resp. realizability interpretation;

not necessarily for that intended by Brouwer and Heyting. In particular, the original work left open whether — for appropriate translations of t_d^G and t_d^K into the language of arithmetic — $A\,[x/t_d^G]$ or $A\,[x/t_d^K]$ or both are formally derivable, which would of course ensure that they hold for the intended meaning.

Incidentally, the formulation above gives a concrete purpose to Gödel's warning on p. 286 of (13) against confusing his interpretation and the orthodox meaning. The warning serves also, at least indirectly, as a correction of his blunder about *eine gering abweichende Interpretation* in (5), pointed out in 2a (i).

(ii) Without emphasizing the issue in (i), Kleene soon found a variant, socalled q-realizability, s.t. the resulting

$A\,[x/t_d^{qK}]$ is not only q-realized, but formally derivable.

The translation of t_d^{qK} into arithmetic language depends essentially on the particular coding of partial recursive functions used (in q-realization).

Gödel's scheme has not been modified equally simply, least of all by him who saw in such work only *Kleinarbeit*, even when done by others...

Since the work reported in 3c(i) — on the stability of E-theorems, in other words, on the equivalence of the various operators μ_0 — the whole matter is moot:

t^G, t_d^K, t_d^{qK} and many more are *equal up to conversion*,

tacitly, by suitable rules and for suitable numberings of the partial recursive functions in the case of realizabilities. — On the other hand,

(iii) Gödel's scheme remains more convenient than Kleene's if not merely some realization of provable $\exists xA$, but an idea of the class of socalled provably (total) recursive functions is wanted. Gödel's schemata for primitive recursive functions provide an elegant description of the class of functions provably total recursive in Heyting's (or classical) arithmetic, and thus a mathematically memorable, not only accessible, instance of a formally undecided $\forall\exists$ sentence: expressing the computability of primitive recursive terms.

Digression for specialists. First, at least so far the description of that class extracted from Gentzen's analysis, in terms of α-recursion for $\alpha < \varepsilon_0$, has been far more useful, for combinatorial and number-theoretic problems about rapidly growing bounding functions, than Gödel's scheme of higher types. (Below we shall return to his soft spot for these objects.) Secondly — and less trivially —, there seems to be a genuine obstacle to modifying Kleene's scheme inasmuch as the use of partial functions, as opposed to using *all* total recursive functions, is necessary for realizing the laws of i.l., even of its propositional part.

Where does one find suitable proper subclasses?

of the class of partial recursive functions, retaining their most highly advertized virtue: a *universal element* that enumerates the subclass from a few initial functions. — For superspecialists: all this applies to functions and functionals of lowest type. Now, Kleene's scheme S9 expresses auto-enumeration but S1-S9 do not generate all partial recursive objects when applied to the principal classes of operations of higher type, for example, the countable functionals (beyond the lowest type). Other, comparably elusive differences between the lowest and higher types will come up in (e) below in connection with bar recursion.

(iv) A most striking difference between Gödel's scheme and Kleene's realizability concerns negative formulae, say F^- (except for $\forall$ formulae, which are left uninterpreted by both). The version of realizability in [T], p. 214 (i) has, literally, *nothing* as the only possible realization for any such F^-, and so extracts no information (except realizability); neither from F^- nor even from a proof of F^-. In contrast, for elementary A and B, Gödel's scheme treats the incidentally very common negative formulae $\forall x A \rightarrow \forall y \rightarrow \forall z \rightarrow B$ and $\forall y \rightarrow \forall x \forall z \neg (A \rightarrow B)$ like $\forall y \exists x \exists z\, (A \rightarrow B)$, thus contributing to a principal concern of mathematics: of unwinding prima facie non-constructive proofs of $\forall\exists$ theorems.

Naturally, often the additional information supplied by Gödel's scheme is not needed, for example, for many crude provability results or, as mentioned repeatedly in this article, for *E*-theorems (of intuitionistic systems). Then realizability is more efficient than Gödel's scheme or the no-counter example-interpretation, let alone, proof-theoretic methods.

(v) Gödel made a point of warning me that he had not given any thought to the objects meant by (his) terms of finite type. The only interpretation he had in mind was *formal*, as computation rules obtained when the equations are read from left to right. Gödel had the impression, in 1955, that ordinals $< \varepsilon_0$ could be assigned to those terms so that each computation step reduced the ordinal. But this was done only much later by Howard [H1], and then more elegantly in [H2].

For reference below concerning a comedy of errors: at the time I did not listen to Gödel's warning since I knew how I was going to understand his terms. The key words are: recursiveness and continuity, the two pillars of the constructive part — as understood in the mathematical tradition — of algebra, resp. topology.

(b) *Background* (ctd). A few minutes after that first conversation in (a), we found ourselves waiting for the Institute bus that took us to the other end of Princeton where Gödel — and, at the time, also I — lived. He added a *caveat emptor* about the

> Aussichtslosigkeit, that is, hopelessness of doing anything decisive in foundations by means of mathematical logic

generally, and by use of the ideas he had just talked about in particular. (One might trick intuitionists into believing that his scheme was constructive.) Again I did not pay much attention, in accordance with my expectations of foundations already at that time.

In 1954, at a congress, I had described foundational 'issues' — there, in connection with finitist proofs — by: one man's meat is another man's poison. And soon I was going to describe (my) interests in such matters as a 'calculated risk'. As mentioned already, for example, in Note 3a(ii), my reservations differed from the more familiar variety since I saw no logical defects such as intrinsic lack of precision; cf. the introduction to [BP] for a different formulation of an equivalent view of my interests.

As for decisiveness there are two complementary points. Unlike Gödel I never expected foundations to do better than ordinary science with questions like (his favorite): What is the world really like? cf. section 4 of [Wi]. What I have always expected — and continue to expect — from logic is a correction of various naive conceptions or aims, partly enshrined in the foundational literature. There need be nothing indecisive about such corrections; for example, to take a familiar parallel, about correcting the aim of astrology to predict human destiny from the position of the planets (rather than predict their orbits); cf. Note 4 on the aims of digital intelligence. In fact, no ordinary scientific result can be quite as decisive or final; even if it is right as it stands, there are usually better new questions. And when the refuted aims have popular appeal, as they often do, the refutations have a wide market; cf. the peroration of [RS].

I believe, the main lesson I have learnt in the last 30 years — incidentally, *after* having tried the opposite scheme of pursuing pedantic distinctions, as in the appendices of [KK] — is this. Remarkably often the defects are so elementary that a

bon mot can be the appropriate literary form

of a refutation. In such cases — not in all! cf. the end of [RS] above — the ritual of (the literary forms of) mathematical logic simply distorts the 'epistemological situation'.

(c) *1955-1957:* background to Gödel's (13). All the results listed below are stated in or are corollaries of Troelstra's compendium SLN 344 [T]. But a selection is needed to present (13) as the gem it still seems (to me). By (f) below this is not Gödel's own selection.

(i) Various classes of functions of finite type were described that both fit the general idea of constructivity and satisfy the axioms in (13) for primitive recursive functions. No 'reduction' is involved since those functions are defined in arithmetic terms, for example, in the case of HEO, or, in the case of the countable functions, in the language of second order arithmetic. But the axioms needed to prove closure under primitive recursion are conservative over first order arithmetic.

This kind of work was enough for various formal independence results and for describing the functions and functionals defined by Gödel's schemata in familiar, ordinal-theoretic terms.

(ii) To avoid the impression of absurdity created by using functions of finite type to interpret mere arithmetic — parallel to proving the consistency of ω-induction by ε_0-induction; but cf. also the end of Note 5 on Spector's education — the emphasis was shifted from Heyting's arithmetic HA in (13) to HA^ω, its by now familiar extension to finite types, with or without various forms of choice, already mentioned in 2a.

The most memorable result, described as 'principal' from the start, is the *classical* equivalence

$$(*) \qquad A \Leftrightarrow \exists s \forall t A_0$$

where A is formulated in the fragment $(\neg, \wedge, \forall)$, $\exists s \forall t A_0$ is its interpretation according to Gödel's scheme, s and t range over the countable functionals, and s may be required to be *recursive*. The proof uses so little that it applies literally also to arithmetic A, with s and t ranging over HEO.

(iii) Practically all crude questions about (ii) were settled; for example, functions s and t of *bounded* type are not enough. Also for A in the language of second order arithmetic, (*) need not hold when s and t range over HEO nor when s is defined by Kleene's schemata S1-S9 in a (iii) above, and t ranges — as required originally by Kleene — over arbitrary functions. — *Remark* (verified by U. Berger). Since the schemata define dense bases for each countable type, (*) can be sharpened: for countable t, an s is defined by S1-S9 applied to countable arguments. Actually, S9 can here be replaced by μ-recursion (in A_0).

(iv) As to *theorems A* of some formal system or other, two points were clearly recognized.

By general theory (*) can be sharpened; the range of s can be restricted to some r.e. subset of the range in (*).

By inspection of mathematical practice, most analytic theorems can be proved in 'weak' systems that are conservative over arithmetic anyway, and interpretable by use of primitive recursive functions together with a few auxiliaries (such as the socalled fan theorem functional in d(iii) below). — Admittedly, no really striking new uses of the interpretation were found, comparable to those that the (formally much less appealing) no-counter example-interpretation, which uses only functionals of lowest type, had provided in the preceding decade.

Principal omission: by a fluke, HRO was overlooked, which is the analogue to HEO when extensionality is dropped.

I reported on these matters in 1957, both at Cornell and at Amsterdam. There was evident, fairly general interest. In 1958 an opportunity presented itself to Gödel to give his own exposition, including second thoughts about the work he had started in the forties. In effect, but perhaps also by intention, there was remarkably little overlap with what I had said in my talks and with what others, for example, Kleene had said about realizability.

In particular, and in contrast to his original work, a main stress was on a primitive notion of effective rule of finite type without extensionality. Under Church's thesis this reduces to HRO, the object that had been overlooked.

(d) Gödel's last full-fledged paper. The opportunity referred to above was a Festschrift for Bernays' 70th birthday. Bernays had been associated with Hilbert's program on finitist foundations, and it might be added that Hilbert himself had already introduced schemata of which Gödel's are a special case; incidentally, without discussion, as if they were obviously finitist in the sense he meant.[4] It is hard to think of a better stage for Gödel's exposition.

We shall here concentrate on the principal novelty — both absolutely speaking, and to me personally when I saw (13) in 1958 —, the primitive notion of effective rule already mentioned at the end of (c). Gödel had never breathed a word to me about his project of exploiting such a notion.

In retrospect Gödel's step is seen to fit very well his philosophy in set theory where he expected wonders from — working with and — brooding over the (primitive) notion: subset - of. Similar wonders could then be expected in the area of constructive mathematics from similar attention to the notion of effective rule. And the emphasis on higher types here fits his faith in higher types in set theory expressed in axioms of infinity; cf. p. 205 of [RS].

As to the formal details of (13), most of them are superseded by the later literature, which had to correct some oversights. For the record, I still find the paper agreeable to read. When this came up in conversation Gödel replied: No wonder (kein Kunststück), there are no proofs. But this alone would not make a gem.

Today, after more than 25 years, I regard (13) as a most artistic package of a jumble of ideas, some of which will now be explained.

[4] Cf. the sketch for a collection of sets that satisfy CH, at the end of *Über das Unendliche*. Hilbert's ground type consists of all (constructive), not only the finite ordinals. — Presumably, the subject of finitist rules (and possibly even of proofs) would become a little more rewarding if the following distinction adumbrated in (13) were pursued; cf. the progress in section 3 by distinguishing between 3 kinds of systems for which rules are intended to be effective. The finitist literature refers both to *finiteness* and to *visualization* (in German, Anschauung). The idea of a hereditary finite operation — without restriction on proofs — is developed successfully in recursion theory. The idea of visualization derives from geometry, and is — without exaggeration — at the opposite extreme from finite, tacitly, discrete mathematics, at least, as these things present themselves. (This may change when more is known about human data processing.)

(i) Asymmetry between rules and — the ranges of — their arguments. One feature that Gödel emphasized increasingly in conversations during the decade after (13) appeared, was the possibility of exploiting the amorphous character — or, if preferred, our ignorance — of the totality of all effective rules. More fully, a rule is accepted only if it is understood to be well defined for all effective arguments (of appropriate type), even though little is — or can be — known about this possibly growing totality. This situation is only superficially paradoxical, to adapt the wording of footnote 1 on p. 283 of (13) about propositional and other logical operators — for the class of propositions — meant by Brouwer and Heyting. There are two obvious illustrations from related areas.

First, ignorance-in-principle is an effective source of knowledge in the theory of lawless sequences, $\neg\forall x\neg\,[\alpha(x) = 0]$ is an immediate consequence of α being given by a finite initial segment. But also

$$\forall\alpha\forall\beta\,[\forall x\,(\alpha x = \beta x) \vee \neg\forall x\,(\alpha x = \beta x)]$$

is seen this way; α and β are either given as identical objects or it is impossible to prove $\forall x\,(\alpha x = \beta x)$.

Secondly, literal models of the asymmetry envisaged are used in that version of recursive analysis which requires its *recursive operations* to be defined on *arbitrary reals*, and not only — in the Russian tradition — on the recursive reals. Closer to home the recursively countable functions of type $\sigma\mapsto\tau$ are defined on arbitrary countable functions of type σ; cf. also e (iii) on the relevance of all this to bar recursion.

It will not have escaped the reader's notice that quantifier free systems like Gödel's T in (13) have two obvious, obviously different interpretations: the variables may range over all effective rules or simply over those that have been recognized-in-principle as effective; in other words over the closure under various elementary operations of the class of those that have been literally recognized as effective. A more delicate point concerning the constants will come up in (ii) below.

Remark. The existence of the literal models in familiar terms mentioned in the penultimate paragraph evidently reduces one's expectations of miracles from the primitive notion; cf. Note 6 and 2d (v) *in fine* on natural sense.

(ii) Definitional and demonstrable equality between terms, possibly containing parameters. There is no mystery here, at least, for those familiar with the literature on *normal forms* (and equality up to renaming variables) in the λ-calculus. Definitional equality is equality of normal forms of those $\lambda x f x$ and $\lambda x g x$ that happen to have normal forms. Though $\lambda x g x = \lambda x g x \Rightarrow \forall x\, (fx = gx)$ the converse is not generally valid, except, of course, in specially concocted extensional models.

The transfer of those ideas to typed systems like T is evident, especially if one goes back to Gödel's original formal interpretation in 4a (iv) in terms of computation rules. This *minimal* definitional equality relation for models of T was examined by Tait [Ta]. Here every term has a normal form — in contrast to the λ-calculus —, the equality relation is recursive, but not provably recursive in formal arithmetic. His proof uses the machinery of arithmetic (and more) in his definition of hereditary computability, and not inspection of any primitive notion of effective rule of finite type; in *sharp contrast to sets* where, for example, Zermelo's axioms are verified on sight for all limit ordinals from a description of segments of the cumulative hierarchy; cf. the remark at the end of d (i) and Gödel's expectations from the primitive notion of effective rules.

The literature seems to have neglected non-minimal equality relations where the constants are interpreted by rules that permit other reduction or computation steps besides those explicitly included among the (equational) axioms (when read as computation rules from left to right). This is the 'delicate point' near the end of (i).

In (13) Gödel mentions definitional equality between terms of type different from 0, but does not get to grips with it. The context of (13), a functional interpretation of arithmetic, is not very suitable since the interpretation does not contain such equations. Of course, the matter arises if HA^{ω} is extended by equations with a corresponding functional interpretation along the lines of c (ii). — In contrast, axioms of extensionality can be formulated already in the old context, but require rather different functionals for their interpretation; in a sense made precise by Howard on pp. 454-461 of SLN 344. A further contrast is found in this area between the axiom and the rule of extensionality, the latter being satisfied by the primitive recursive functions and many other such classes.

(iii) The last sentence of (13) states, without comment, that the fan theorem is interpretable. This requires the socalled fan theorem functional or, more simply, a modulus of continuity for all functionals Φ of lowest type, that is $(0 \mapsto 0) \mapsto 0$, applied to all functions bounded by f (of type $0 \mapsto 0$). The fan theorem is certainly evidently interpretable by a recursively countable functional, and equally evidently not interpretable in HEO or HRO. An argument is needed to show that it is not interpretable by an object generated from the countable functionals by Kleene's schemata S1-S9; cf. [GH].

However, it is not at all evident that the fan theorem is interpreted by an effective rule of the kind considered in (13); equivalently: whether there is an effective modulus of uniform continuity for effective Φ and f. By the end of (c), under Church's thesis it is not so interpretable since then the effective rules are those of HRO.

In terms of (i) above, ignorance could be a source of knowledge. Specifically, if we know sufficiently little about the totality of effective functions (of type $0 \mapsto 0$) then Φ can be recognized to be effective only if it is also recognized to be continuous (for the product topology). And then we also have a modulus of uniform continuity. — *Reminder.* The schemata of Gödel's T are not explicitly required to be continuous (nor to be applied only to continuous arguments). But for any primitive recursive Φ there is also a primitive recursive M_Φ of type $(0 \mapsto 0) \mapsto 0$ that satisfies demonstrably the requirements of a uniform modulus of continuity

$$\forall g \forall g' \{[\forall x \leqslant M_\Phi(f)]\ (gx = g'x) \wedge \forall x\, [(gx \leqslant fx \wedge g'x \leqslant fx)] \Rightarrow \\ \Rightarrow \Phi\ (g) = \Phi\ (g')\}.$$

A good deal more was said on related matters in conversations with Gödel. But it is better left for the digression in (e) below.

What was not said — partly because I had not cottoned on to the parallel between the primitive notions of set and of effective rule — is more interesting. As matters stand the parallel holds out little hope for effective rules; cf. top of p. 213 in [RS]. The last 20 years have shown that our knowledge of sets in segments of the cumulative hierarchy, for example, as expressed in current axioms of set theory, is simply more rewarding when applied to suitably defined structures, in particular models of set theory that satisfy additional conditions. We literally know more about the constructible sets: they demonstrably satisfy the current axioms, but also

GCH while we do not know whether the cumulative hierarchy satisfies CH, nor even whether their second order version decides GCH; cf. Note 6. Exaggerating very little, the primitive notion of set serves to answer the question (if one insists on an answer):

from which stock of sets are the constructibles defined?

Practically, most problems about the constructibles are not very sensitive to the answer (but depend only on the stock satisfying certain closure conditions). But if one wants to commit oneself to some universe of objects the primitive notion gives one the means at a small price; this responds to the worries at the end of App. 2A of [KK].

Similarly, in the constructive theory of functionals we have Brouwer's inductive definitions of the type $(0 \mapsto 0) \mapsto 0$; cf. the introduction and the appendix of [KT2]. From which stock of functions of that type are they selected? Gödel's primitive notion advertized in (13) is a good choice. In conclusion, there is no mystery about definitional equality; but there also is not much prospect for any spectacular uses of that notion.

(e) *Bar recursion* or, more soberly, recursion on well founded trees became prominent in connection with [S], when Spector used it to describe the provably recursive functions and functionals of lowest type for *formal* classical analysis. This sharpens not only (*) of 4c (ii), as envisaged in 4c (iv), but also the remark in 4c (iii) since Spector's bar recursion is definable by S1-S9 on the countable functionals. As will be recalled, his version formally extends Brouwer's original bar recursion for decidable trees labelled by objects of a decidable species, for example, the natural numbers, to trees labelled by objects of higher type. At the time the burning question was: Which objects?

Starting with Gödel's (13) the answer is: effective rules. — As is obvious from my footnotes to [S], written some 3 years after (13) had appeared, I had remained attached to the countable functionals (as labels). It is time to look at the other answer. — Independently of all ideology of i.l. concerning the evidence of bar recursion, for example, in Gödel's banter reported in Note 5d (v), there is a question of 'raw' interest, broached already in the digression at the end of 4a (iii) about differences between functionals of lowest and higher types, specifically, between $(0 \mapsto 0) \mapsto 0$ and

$(0 \mapsto \sigma) \mapsto 0$ for $\sigma \neq 0$; more pedantically, it concerns contexts in which rewarding differences occur, and concepts in terms of which they can be stated. There is certainly no lack of candidates! For example, $0 \mapsto 0$ is not, $0 \mapsto \sigma$ is isomorphic to 0, resp. to σ (unless the functions $0 \mapsto 0$ considered are effectively enumerable); decidability of the type 0, in other words, of natural numbers, is not, that of σ is problematic. Perhaps most pertinent to what Brouwer first called a proof of the bar theorem is the possibility of an *inductive definition* of the species of well-founded trees labelled by objects of type 0 (and, admittedly, $0 \mapsto 0$) in the penultimate paragraph of 4d; that species is demonstrably closed w.r.t. all operations that come to mind. The literal analogue is simply not true for higher types; at the very least, one would need some new idea about the notion of *freely chosen path* involved in well foundedness.

This digression will now be concluded by odd formal facts that *seem* to me relevant here, as I mentioned to Gödel when the occasion arose. He explicitly rejected them, and, it might be added, fully in accordance with his expectations of wonders to be seen by looking at the primitive notion of effective rule (undistracted by the *Kleinarbeit* that goes into (i)-(iii) below).

(i) This item concerns the passage to higher types in the case of the bar theorem. Realistically speaking, Brouwer's idea of a 'fully analysed' proof — specifically, of $\forall\alpha\exists xR(\alpha, x)$ for decidable R — is not compelling when taken literally. If R is $\alpha[\alpha(0)] = x$ then

$\forall\alpha\exists xR(\alpha, x)$ is evident without further analysis, but not fully analysed

in his sense. However, there is a perfectly good formal analogue, Gentzen's notion of (infinitary) cut-free proof; at least one of its versions specializes to Brouwer's notions for the proofs considered. As already Aristotle knew, proofs using modus ponens need not be further 'analysed' — by cut elimination — to be convincing.

When passing to higher types α^σ, of type $0 \mapsto \sigma$, the question is: what is now the appropriate notion of cut-free proof, and do at least the usual principles admit cut elimination? If the α^σ are *defined* objects, say, neighbourhood functions α of countable α^σ, thus satisfying a suitable (analytic) condition C^σ then it seems open whether a cut-free proof of $\forall\alpha\,[C^\sigma(\alpha) \to \exists xR(\alpha, x)]$ will generally look at all like Brouwer's fully analysed proofs. — Here it is understood that the 'usual' principles are meant to include not only continuity, but also other mathematical axioms, especially those that concern the generally lawlike *data* for the choice sequences considered.

Digression (especially for readers of the digression in 4a (iii) above). Provided non-ideological differences between functionals of lowest and higher types are formulated imaginatively, their examinaition need not be mere Kleinarbeit. On the contrary it could provide an exception to a general philosophical insight (and would thus be an insight, too). Specifically, at least usually, formal differences between instances of the same scheme — differing in logical or type complexity — *falsify the epistemological situation.* For example, as proof theory has shown, instances of induction of different logical complexity have different proof-theoretic strengths, even though they all derive their evidence from one and the same principle. Without exaggeration, the whole matter of proof-theoretic strength is an artifact w.r.t. to the evidence of proofs. However, while generally valid the insight leaves open the possibility of *discovering* situations where the formal differences are relevant, in which case the latter may fairly be said to have helped in the discovery. In the particular cases of bar induction of lowest and all finite types we start with the formal knowledge of their different proof-theoretic strenghts; roughly, of Π^1_1 -CA and Π^1_∞-CA. By Note 8d (iv) there is a chance of a reinterpretation in non-ideological terms; by reference to proofs (which are constructions, even when the proofs are non-constructive). In any case, *some* reinterpretation is needed to get away from the socalled consistency problem of classical analysis, which rests on highly dubious doubts. — Here ends the digression.

In summary, of course the formal analogues do not settle Brouwer's claims about arbitrary (convincing) proofs. But in view of 2b, and of experience in set theory already cited — on p. 213 of [RS] and above, — arbitrary proofs of $\forall\alpha\, \exists x R\, (\alpha^\sigma, x)$ may be less rewarding. Trivially, all this applies mutatis mutandis to the primitive notion of effective rule and defined models more or less inspired by it; cf. Note 6, but also the end of 5d (i) and 5d above.

Bibliographical remark. Recently, several more models have appeared, for example, by Bezem [Be] that seem rewarding, even if not necessarily for the properties generally emphasized (by logicians). After all, Gödel's incompleteness theorem is rewarding enough though certainly not for — what logicians consider to be the great 'mathematical' discovery of — the fix point 'lemma'.

(ii) This item concerns a surely noteworthy if not often noted aspect of Spector's proof, quite independent of agonizing re-appraisals of the principles used. The proof has obvious mathematical wit, and so there is surely something behind it. True enough; but certainly not the result stated. Specifically, for example, by [HK], classical analysis, without any choice and with the axiom of dependent choices, has easy reformulations in terms of bar induction of type 0, resp. 1. Howard [H] has given straightforward interpretations of those reformulations by use of bar recursion of Spector's type.

So it is a problem to discover a context where the combinatorial wit in Spector's original proof is actually relevant.

(iii) The last item, like the end of (i) above, also involves a parallel with — Gödel's experience in — set theory, but a more formal oversight. It is the role of higher types, with the first few steps illustrated already in (13): from primitive recursive functions of lowest type to all finite types or, equivalently, in ordinal-theoretic terms, from ω^ω to ε_0. There are 2 points.

For the general context of i.l., as already mentioned, — this kind of — proof theoretic strength goes with algorithmic inadequacy.

But even judged only for such strength, the parallel is deceptive, because in set theory higher types derive their strength from closure under the power set construction; for example, without the latter, models of replacement, which pushes up the types, can be defined by use of comprehension. In the context of i.l. (that Gödel had in mind) one does not have any analogue for the power set. For 'straight' limitations of higher types in i.l. — not relying on analogies with set theory — cf. the autonomous progression in Problem 3 on p. 156 of [K5], and H. Friedman's models in Π^1_1-*CA* that serve to solve it.

(f) 'Not with a bang' describes Gödel's last attempt — in those notes to (13) left in the Nachlaß, and referred to already — to squeeze out results of cosmic significance from (13). Two droplets will convey the flavour.

(i) Proofs represented in T are claimed to be *analytic* in the sense of Kant; in other words, they use only — properties of — concepts implicit in those used to state the theorem proved. Now, proofs of Π_1^0 theorems, expressing the insolubility of diophantine equations, demonstrably may contain (equational) axioms for primitive recursive functions of *unbounded* type. If the latter are implicit in + and × what is not? but cf. a distinction between analytic proofs and analytic axioms below.

Overlooking the distinction may have been a mere oversight. The *philosophical* sin (of omission), at least, in the sense of section 1, is that Gödel does not even begin to examine the relevance of Kant's ideal or, more generally, his question: How are proofs possible? cf. bottom of p. 162 of [RS]. Incidentally, as has often been observed, Kant's ethereal ideal corresponds to the venerable tradition of *purity of method* in mathematics (cf. p. 163 of [RS]; in agriculture such purity is required already in the Old Testament, Deutoronomy 22, 9-11).

The last paragraph is not merely irreverent. It suggests an examination of Kant's ideal by reference to the *whole body of mathematical experience*, where purity of method has been pursued since the Greeks, and its defects, especially, the notorious loss of Beziehungsreichtum, have become apparent. Without exaggeration, the problem is to discover corners where the principle of purity of method is appropriate, for example, in connection with the extraction of algorithmic information from proofs of ∀∃ theorems; cf. the end of 3c (i) on the marginal character of algorithmic aspects of proofs.

A warning about Gödel's notion of *analytic axiom* (in the second of the remarks from the seventies cited in footnote 2). Here axioms of infinity in set theory are explicitly regarded as analytic in the sense of 'explicating' the concepts occurring in them; as if not every property of a concept contributed to its 'explication'. In this case logical deductions from analytic axioms, for example, of the insolubility of a diophantine equation from axioms of infinity, are not generally analytic proofs (respecting purity of method); cf. bottom of p. 197 of [RS].

The next item involves, I believe, only an oversight.

(ii) In that same collection of notes Gödel suggests that some kind of reduction is achieved by the decidability of definitional equality. Though the words are vague, the meaning seems plain enough. He apparently forgot that the converse to $\lambda x f x = \lambda x g x \Rightarrow \forall x\, (fx = gx)$ is not valid; cf. the beginning of 4d (ii). But even if this particular impression is wrong the following is surely generally right.

The wish to draw conclusions of cosmic significance is as sure a way to make mistakes as the kind of lack of interest that Note 5a sees behind the error in the last sentence of (8) about adding = to the socalled Gödel case of $\forall^n \exists^2 \forall^m$ formulae.

To end on a more 'constructive' note: (13) — and even (i) and (ii) above — can serve for contrast; cf. Note 6c.

(g) *Another view of* (13) itself, even without considering all the work that refers to it.

(i) It is a memorable reminder, using a minimum of scientific experience, of the potential of higher types in constructive mathematics. As the title of (13) stresses, this aspect had been neglected. Of course, the no-counterexample-interpretation also used higher types (and a paper was devoted to the need for something of this sort [K2]). But the role is not nearly as memorable as in the scheme of (13) where the *only* formal difference from ordinary primitive recursive arithmetic is the use of higher types.

(ii) As already mentioned in passing in 4a (ii) and in line with (i) above, computability of — even the purely numerical — terms is not derivable in formal arithmetic. Since the rules of T are memorable — to repeat from 4 a (iii), not merely accessible —, (13) provides a memorable formally independent sentence in the mathematics of computation. And it is useful, if regarded as a *warning* concerning the algorithmic inefficiency of such schemes as that of (13); cf. 4b on Gödel's concern in 1955 about 'tricking' intuitionists. Admittedly, it is not as memorable as his second theorem in the branch of mathematics called 'metamathematics'.

(iii) Whatever its scientific value the notion of effective rule with definitional equality is so close to the surface of our logical subconscious that, as a matter of logical hygiene, it is salutary to take a look at it. (13) — so it still seems to me — helps us do this in a remarkably painless way, compared, for example, to generalized recursion theory of the seventies; cf. the review of [F].

In b (iii) of Appendix 1 there is a sober view of the incompleteness theorem. It is related to, say, the business of mind and matter as (g) here is related to (f) above.

APPENDIX 1

GÖDEL'S EXPOSITORY STYLE IN THE THIRTIES: A NEGLECTED ASPECT.

He gave *increasing* attention to intuitionistic — or, more precisely, to the even stricter finitistic — requirements on metamathematical methods; specifically, in connection with (a) the completeness of predicate logic, (b) the incompleteness of formal systems for arithmetic, and (c) the consistency of the axiom of choice and the generalized continuum hypothesis relative to the usual axioms of set theory.

Our concern here is only the extent to which those requirements are rewarding for (a)-(c); both in the long run, and for short term effects. The latter were always Gödel's main concern, at least, in conversations on titles of papers and on terminology; cf. Note 7 (but recall also Note 6 and earlier parts of this article). — It is beyond the scope of this paper to go into the delicate relations between the facts of Gödel's style here considered, and his later views or memories of them. Of course, those relations are fascinating, and may even be rewarding when treated in an inspired way. But only the most coarse-minded among us could be tempted to speculate on such matters in terms of ordinary knowledge about them, cf. also the end of Appendix 2.

In summary, Gödel's discussion of (a), especially in his dissertation,[5] was remarkably penetrating and explicit, albeit a little clumsy by current standards; he was much less explicit about (b) at the start, but added pertinent improvements later, such as those in 2b (i); almost perversely pointless with (c), with comic consequences related in (b) on p. 197 of [RS].

(a) The question of *completeness* of (Frege's) rules for predicate logic, is as old as the hills; that is, whether for all logical formulae *F*

[5] *Correction.* In [K 8] I described Gentzen's patience as unusual among gifted young logicians, tacitly among contemporaries. I had in mind principally Herbrand's free associations and Gödel's terse publications. At that time I did not know Gödel's dissertation.

$$(*) \qquad \mathrm{Val}\, F \Rightarrow \vdash F$$

where Val means (classical) validity in arbitrary structures and $\vdash$ means formal derivability. Brouwer raised it in his dissertation, without however stopping to *paraphrase* (*) so that it becomes even a candidate for being settled by methods of i.l. Hilbert's paraphrase is described on p. 178 of [RS].

In the introduction to his dissertation Gödel considers

$$(**) \qquad (\vdash F) \vee \neg \mathrm{Val}\,(F)\,,$$

which is classically equivalent to (*), and derives *decidability* of $\vdash$ from any proof of, tacitly, a paraphrase of (**) in i.l.; here the logical symbols $\vee$ and $\neg$ mean disjunction and negation of i.l., and Val (F) is 'soundly' paraphrased by, say $\ll \mathrm{Val}\,(F) \gg$; in particular, $(\vdash F) \Rightarrow\, \ll \mathrm{Val}\,(F) \gg$. — Reminder. $\vdash$ being r.e., a proof of $(\vdash F) \vee \neg \ll \mathrm{Val}\,(F) \gg$ in i.l. would provide, for each F, an object d_F s.t.

d_F is a formal derivation of F (a decidable matter) or $\neg \ll \mathrm{Val}\,(F) \gg$, and hence, by soundness, $\neg \vdash F$.

It was a brave attempt. But the intuitionistically invalid switch from (*) to (**) would exclude also intuitionistic completeness proofs for undecidable systems of i.l.; cf. 2a. — However, by p. 169 and 178 of [RS], the idea above is enough to derive decidability of $\vdash$ for Hilbert's paraphrase of Val (F).

An historical titbit: an objection not foreseen in the dissertation. According to Mostowski, in a conversation in Tarski's presence, the latter and his students had no confidence in Gödel's paper when they saw the relevant issue of the *Mhfte* in Warsaw. Why? Gödel had not formally defined validity! Anybody who is surprised by this knows *ipso facto* that he simply has no feeling for the subject. (I had the good luck more than a quarter of a century later to experience the reactions of clever people to completeness proofs for i.l. in the fifties, and incompleteness proofs in the sixties, also without formal definition of — here, intuitionistic — validity; key word: basis results.) Of course, when Tarski met Gödel in Vienna soon afterwards, confidence was established. In the other case personal explanations, sometimes repeated, verbatim, half a dozen times, also helped to establish confidence (of some people); cf. Appendix 2 for other episodes of this sort.

Evolution of a perspective on the completeness theorem; cf. the last paragraph but one on p. 160 of [RS] on the relative glamour that the completeness theorem had when it appeared, and would have had 50 years earlier. Here we do not go into such a general perspective, but only into aspects to which i.l. is relevant; both for the wider meaning of 'constructive' in the sense of definable and the stricter intuitionistic or even finitist sense concerned with provability. (The two senses were elaborated at the beginning of this article.)

(i) Gödel's result that Hilbert's paraphrase of (*) is false has some anti-ideological consequences. The requirements of i.l. are sterile in connection with (*) because the conjectures, expressed by paraphrases that come to mind easily, are simply false. In the meantime there has been a shift of emphasis, in terms of (the absence of) recursive models; cf. p. 180 of [BP], written incidentally without my having seen Gödel's dissertation. He never told me about his anticipation of the general idea 30 years earlier; cf. the oftquoted Note 2a.

(ii) The particular rules set down by Frege are simply not adequate for the study of delicate aspects of logical proofs; consequently, claims about such aspects in terms of those particular rules are merely pretentious, cf. p. 187 of [RS]. So it is — philosophical! — progress to *discover* questions with answers that are less out of proportion with what we know of logical phenomena. The shift in (i) above leads to such questions in terms of *recursion-theoretic complexity;* of models, and, more generally, of sets of valid theorems. This is a definability property, and thus a concern of i.l., at least in the weak sense. As a refinement, we have:

(iii) Distinctions between total and semivalutations and their recursion-theoretic complexities lead to analysis of different proofs of completeness.

Bibliographical remarks. Shoenfield's term 'characterization' instead of 'completeness' theorem is obviously intended to convey (ii), but has not caught on. Hasenjäger [Ha] used (iii) to analyse convincing differences between Gödel's and Henkin's proofs of completeness. Both points are elaborated in [KMS], and more recently in the review of [SV].

A broader perspective, especially on paraphrases, is developed in Appendix 2.

(b) Adequate background on the *incompleteness* results about systems for arithmetic is to be found on pp. 170-176 of [RS], complemented by section 2b above. Naturally, also for i.l. the formulation of incompleteness w.r.t. sentences expressing the soundness of Π_1^0-theorems is superior to Gödel's original, Hilbert-style formulation in terms of consistency; cf. 2b (i). As also mentioned in 2b (i), the purely 'syntactic' reformulation of completeness for arithmetic, favoured by Hilbert, namely

(*) for all closed formula A, $(\vdash A) \vee (\vdash \neg A)$

would not be expected in i.l.

In connection with Hilbert's program it was necessary to prove incompleteness by finitistically acceptable methods, and Gödel emphasized that he had done so. (This is almost, but not quite true; cf. (ii) below.) Inasmuch as he achieved this, his metamathematical methods are valid for i.l., too. Today it seems appropriate to go into some fine points he — and others — neglected at the time.

(i) The *weakness of negation* in i.l. is notorious. For example, a refutation of the Π_1^0 statement (*) above need not furnish a counter example, say A_G: $(\neg \vdash A_G) \wedge \neg (\vdash \neg A_G)$. On the other hand, since (*) is not required by completeness it is not immediate that a counterexample implies incompleteness. But this follows from a little

Exercise. Write Comp A for: $A \to \vdash A$, in other words, completeness for A. Then a counter example A_G, implies: $\neg$ [(CompA_G) $\wedge$ (Comp $\neg A_G$)]. Hint: Since $[(\neg \vdash A) \wedge \text{Comp}A] \to \neg A$, a counter example together with [(CompA_G) $\wedge$ (Comp $\neg A_G$)] implies $[(\neg A_G) \wedge (\neg\neg A_G)]$ which is absurd.

The matter simplifies for the usual systems, which are demonstrably complete for Σ_1^0 (but not generally for $\neg\neg\Sigma_1^0$) formulae, if a variant $\tilde{A}_G$ is used that expresses literally: $\neg \tilde{A}_G$ *is derivable*. Then $\tilde{A}_G$ is Σ_1^0, and so $\vdash$Comp $(\tilde{A}_G)$. Since $\tilde{A}_G$ is $\vdash \neg \tilde{A}_G$, Comp $(\neg \tilde{A}_G)$ reduces to $\neg\neg \tilde{A}_G$, which is false, but not formally refutable (if the system considered is consistent). Since $\neg\tilde{A}_G$, like A_G, is formally equivalent to *consistency,* the usual formulation in terms of A_G can be recovered; cf. below.

Gödel's work was explicit enough not to need the exercise. In fact, $\neg\mathrm{Comp}A_G$ was proved from the consistency of the system considered. The proof of the implication is even finitist because, for any (proposed) derivation of $\mathrm{Comp}A_G$, not only the inconsistency of the whole system is concluded, but a specific derivation of an inconsistency. This is a case of the familiar sharpening, for quantifier-free A and B, of $\forall xA \to \forall yB$ to $(\forall x \leq \sigma y)\, A \to B$ for suitable σ.

(ii) The independence of Gödel's sentence A_G is more delicate because of the assumption of ω-consistency. Is the proof of

$$(\omega\text{-consistency}) \to \neg \vdash \neg A_G$$

valid for i.l.? It is, for example, by a straight application of the negative translation in 2a. But at the time Gödel overlooked the question.

When I met him in 1955 he brought up the matter. It had bothered him until he noticed a footnote in my paper on the no-counter example-interpretation that gave an explicit finitist interpretation of ω-consistency. Actually, all this is a bit superfluous; for establishing underivability of $\neg A_G$, the most immediate requirement is: soundness of derivable negations of Π^0_1-formulae. At the time I talked of '1-consistency'; cf. Smorynsky's detailed obituary of this notion in [Sm]. (For the record, my silly terminology never came up in conversation with Gödel on such matters; cf. Note 7.)

As already announced at the end of section 4, some sober reminders follow concerning the significance, that is, implications of the incompleteness theorem different from Gödel's business about Mind and Matter.

(iii) In the introduction to his paper on formal incompleteness Gödel related his first theorem to — one version of — the program of Frege and Russell about the structure of mathematical concepts, specifically, in terms of formal systems (as opposed, for example, to second order axiomatization, cf. pp. 164-165 of [RS]). Near the end he warned against relating the second theorem to — again, one version of — Hilbert's program; specifically, concerning the possibilities of purity of method for finitistically formulated problems (as opposed, for example, to the business of a final solution for all foundational problems, which is convincingly refuted already by the first theorem). Contrary to a would-be sophisticated view these matters are not of 'purely historical interest', every bright teenager being interested in them. But they do not constitute a principal interest of the incompleteness theorems. This has been compared to Pythagoras' program expressed in the slogan: (rational) numbers are the measure of all things, and the irrationality of $\sqrt{2}$. The latter remains of interest, but not primarily because it refutes such an exaggerated, and, therefore simpleminded program. — Incidentally, 'purely historical interest' is probably better suited to the more difficult works of Archimedes and the bulk of all the excellent mathematics in the 19th century; they are inaccessible to the outsider, and superseded by later results for the specialist; cf. the end of Appendix 2 on other historical matters.

As matters stand today the incompleteness theorems are not literally fundamental discoveries rewarding unlimited elaboration or analysis, but as *samples;* to be compared to the irrationality of $\sqrt{2}$, that is, $n^2 \neq 2m^2$, as a sample of diophantine problems.

(α) The first theorem is a corollary to (recursive) undecidability results about arbitrary Π^0_1 sentences; cf. 162 of [RS]. This was later improved by a variety of incomparable results; on word problems in group theory or on diophantine equations in number theory, and in many other branches of mathematics; cf. the review of [F] and (iv) below.

(β) The first theorem is also a special case of incompleteness of not necessarily formal systems, for example, of such systems extended by all true Π^0_1-sentences; cf. also footnote 2 to 2b (i), and p. 175 of [RS] on implications for the rate of growth of bounding functions in the case of Π^0_2 theorems that are formally independent of all true Π^0_1-sentences.

(γ) Apart from the improved formulations in 2b (i), the second theorem has not only been appropriately extended to non formal systems as in (β), but also to cut-free systems that have been developed since Gödel's original paper; cf. the end of 2b. More usefully,

(δ) As stressed in (b) on p. 175 of [RS], the second theorem serves as a cross check on proposed consistency proofs; cf. also p. 215 of [RS]. This is more useful than it seems just because consistency is so weak (too weak, for example, as a soundness property). Consequently, many metamathematical results, for example, various kinds of normalization, imply consistency formally. Thus the second theorem serves as a cross check on proposed proofs of such results, too.

(ε) The second theorem has been sharpened to conservation results. Thus while the theorem only states that the addition of (the false formula) $\neg \mathrm{Cons} S$ to a consistent system S is consistent, in fact, no new Π^0_1 sentences can be proved in the extended system; cf. top of p. 176 in [RS] for more of the same kind. — *Philosophical corollary*. The interest of this reformulation of the second thorem is independent of dubious, that is, ideological doubts about the legitimacy of, say, current principles S. On the contrary, for such S and the usual interpretation of the formalism, $\neg$ Con S is false, and so the consistency of $S \cup \{\neg \mathrm{Cons}\ S\}$ is genuinely problematic. But the single most satisfactory way available to day for high-lighting the second theorem uses socalled (propositional) provability logic touched in 2b (ii), as follows.

(ζ) The second theorem is the special case of $\Box(\Box p \rightarrow p) \rightarrow \Box p$, Löb's theorem, when $p = \bot$, and Löb's theorem for a system S follows from that special case applied to $S \cup \{p\}$(which is consistent if $\neg\Box p$, and would prove its own consistency if $\Box(\Box p \rightarrow p)$ held for formal derivability $\Box$ in S). Now, as shown by Solovay, together with a couple of pedestrian properties of $\Box$, Löb's theorem axiomatizes completely its propositional theory. In this sense the second theorem is, as it were demonstrably, central for the subject of formal derivability in the usual systems, which have those pedestrian properties.

Certainly, none of (α)-(ζ) is as exciting as (Gödel's) claims about the significance of the incompleteness theorems for the nature of mind and/or matter or as the even more remarkable claims of, say, Hofstadter in connection with digital intelligence; 'remarkable' because a theorem that states what — even perfect — computers cannot do is supposed to provide evidence for the unlimited potential of A.I., which relies on what real time computers can do; cf. section 3 of the main text.

Of course, the incompleteness theorems tells us something of interest about (limitations of) human minds; in particular, how exceptionally gifted people — cf. the second paragraph on p. 150 of [RS] — who talked about the topic (endlessly) could miss such simple proofs. More specifically — cf. p. 161 of [RS] —, how grand models for the structure of mathematics and the laws of thought could be proposed *without any check* on the mathematical properties of those models; cf. Note 1b (ii) concerning Hilbert. Incidentally, this is the way young Gödel saw matters himself. As he once told me, when he submitted the announcement (2), he was prepared (but not hoping) for a publication of those theorems by somebody else before his appeared. In other words he did not think of them as far beneath the surface; cf. p. 154 with his brother's phrase of Gödel 'hiding' his light under a bushel, another way as it were, of expressing Gödel's own view that he had to work very little for those results; cf. Note 3a (iv).

(iv) *A fall out from the Salzburg meeting* concerning Gödel's use of the Chinese remainder theorem. Viewed within the context of his incompleteness paper the attention paid to the language of rings (+ and ×) appears disproportionate; too much for the general problem, not enough for Hilbert's 10th problem; cf. the parallel between formal derivability and solvability of diophantine equations on p. 168 of [RS]. But viewed as part of Gödel's mathematical education, his use is most *satisfaisant pour l'esprit.* As emphasized by Olga Taussky, Gödel followed Furtwängler's lectures on class field theory where the Chinese remainder theorem is used for the very same purpose as in Gödel's paper: to code sequences of elements by a single element. — 40 years later this part of Gödel's paper can be seen as a step towards Matyasevic's result alluded to in (iii)α above.

(c) The introduction to the monograph (10) on the consistency of the axiom of choice and the GCH relative to the remaining axioms of set theory gives pride of place to the *strictly finistist* character of the proof. Several expositions dutifully repeated this emphasis. Objectively, this is surely *le côté le moins intéressant.* As already mentioned in Note 6b(ii), given the potential of (10) and the delay in further work on it, the introduction was hardly effective. For Gödel himself the stress on consistency had comic consequences; cf. p. 197c of [RS].

Around 15 years after (10) first appeared he himself felt ill at ease, and asked me one of his offhand questions about it; cf. Note 3b (ii). If $S \subseteq S'$ and Con $S \rightarrow$ Con S', in other words, if the consistency of S' relative to S is proved in S itself, is there also a finitist consistency proof? The trivial answer is No; for example, if S_1 is consistent and $S = S_1 \cup \{\neg \mathrm{Con} S_1\}$, S is consistent by Gödel's second theorem, and $\mathrm{Con} S \rightarrow \mathrm{Con} S'$ can be proved in S for arbitrary S' since $\neg \mathrm{Con} S$ is provable in S: cf. [K4]. But $\mathrm{Con} S'$ need not even be true. I did not agonize over the proof, and normally I should not have published it. But at the time I was preoccupied with establishing the notion of *conservation,* which I found better adapted to summarizing the interest of then current relative consistency proofs. And the temptation to ask such questions as Gödel's seemed an additional weakness of the notion of relative consistency. Almost 20 years later, in the abstract [JSL 41 (1976) 285-286], I found a better wording of Gödel's question in [K4]. But first some reminders.

(i) Even finitist relative consistency proofs do not assure conservation. *Standard example.* For Rosser sentences R_S, both $\mathrm{Con} S \rightarrow \mathrm{Con}\ (S \cup \{R_S\})$ and $\mathrm{Con} S \rightarrow \mathrm{Con}\ (S \cup \{\neg R_S\})$ have quite elementary proofs; R_S is Π^0_1 and $\neg R_S$ is Σ^0_1; cf. footnote 2a. So the bare fact of relative consistency — and even of its elementary provability — gives no information about conservation.

In contrast, the *inner* model constructions of Gödel or Cohen (preserving $\in$, ω etc.) do give useful conservation results.

(ii) Models — or 'interpretations' in the terminology of the monograph by Tarski - Mostowski - (R.M.) Robinson —, at least, for one finitely axiomatized system (= one formula) in another, immediately yield a finitist relative consistency proof; for example, in Gödel's case of GB and $GB \cup \{V = L\}$. This was, in fact, the point of his introduction. — The converse is false.

Remarks. The only property of 'finitist' used above is that, say, primitive recursive arithmetic is finitist. If not only relative consistency has such a proof, but $\mathrm{Con} S'$ itself (is proved in S) then, by the formalization of the completeness theorem, appropriate models of S' can be defined in S.

(iii) There is a rewording of Gödel's question, with an additional quantitative condition on the relative 'lenghts' of hypothetical proofs of an inconsistency, say $\perp$, in S and S':

(*) $\qquad \forall d'\ [\mathrm{Prov}'\ (d', \perp) \rightarrow (\exists d \leqslant \sigma d')\ \mathrm{Prov}\ (d, \perp)]$

is provable in S, where Prov and Prov' are the proof predicates of S, resp. S', and σ is elementary. Then (*), and hence

$$\mathrm{Con}\,S \rightarrow \mathrm{Con}\,S'$$

has an elementary proof, simply since (*) is Π^0_1, and a Π^0_1 theorem of S can be deduced from Con S by elementary means. This was done in the sequel to the abstract [K4] on Gödel's original question, mentioned earlier.

At the time, (*) was intended as answer to the question on p. 100 of [K11], which summarized what — I thought I or, with luck — we have learnt from Hilbert's second problem: What more do we know if we have an elementary proof of relative consistency?

Again, it turns out that a restriction on the function σ in (*) is critical; less so the method of proof, a proof in S being enough (for an elementary proof); cf. the 'principal element' of i.l. in the second paragraph at the beginning of this article.

(iv) What more do we know from (*) in (iii)? if we restrict σ even further. H. Friedman has given a satisfying answer.

If (*) — of course, not merely relative consistency — is proved in S for an *elementary recursive* σ — of course, not merely elementary in the sense of finitist, in Gödel's off hand question — then a model of S' can be defined in S; with variants when the class of σ is extended. (Roughly speaking, S is replaced by the ramified hierarchy over S of level α, where α is the ordinal usually associated with S; for example, ω^ω when S is primitive recursive arithmetic.)

Naturally, Friedman's result adds nothing to the monograph, which, being an inner construction, produces directly a model with particularly useful conservation properties.

The gushing drivel in [NH], already mentioned in Note 2c and especially 7c(i), obscures Friedman's contribution. Far from being fundamental — except in the sense of being familiar to us since our teens — relative consistency is, on the face of things, even less compelling than consistency. As cannot be repeated too often, the latter is justified by the observation that it is sufficient for the truth of Π^0_1 sentences. (Otherwise, one thinks of consistent liars.) In the case of finitist consistency proofs of S' relative to a dubious S, one thinks of a passage in Mr. Midshipman Easy, where a virtuous wet nurse was to be hired. A girl applied who turned out to be unmarried. Did she not have a child? Yes, but only a very little one. — S' is so little more dubious than S!

Friedman's theorem need not be presented as pursuing mindlessly the ideology of finitist relative consistency proofs. It is also a contribution to philosophy in the sense of section 1; by establishing some significance, that is, some consequence of elementary relative consistency proofs for a sensible purpose; specifically, with a proper meaning of *elementary,* and for the purpose of defining models of S' in S. (His models *may* be rewarding!) Here, finitist would simply be too crude.

Warning. These results involving socalled lengths of derivations should be interpreted as giving significance to the parameters called 'length', at least, in one of the usual senses of significance: the parameters are used to state consequences we want to know about. However, the parameters do not provide any measure of complexity, for example, in the sense of intelligibility, of the proofs 'represented' by the derivations. This is so because the 'representation' is far too crude to serve for any analysis of such delicate phenomena as intelligibility. Specifically, a (possibly quite short) *description* of a (possibly quite long) *formal derivation* is at least as convincing as the latter, and so the 'length' of the latter used above is an artifact in connection with intelligibility.

As for Gödel's increased attention to requirements of i.l. on metamathematical methods, it was not misplaced in (b), but simply fell between two stools in (c); *either* too strong, since only the fact, not the method of proof is relevant to uses of conservation *or* too weak since for the quantitative version (*) of relative consistency subdivisions within finitist mathematics are critical.

APPENDIX 2

SOME CORRESPONDING HIGH SPOTS OF EARLY CLASSICAL AND I.L.

The periods meant are roughly 1880-1930 for classical logic and 1930-1960 for i.l. The 'correspondence' concerns (a) completeness, tacitly, for intended meanings, (b) concocted — or, more reasonably, discovered — meanings, and (c) relevance of intended and concocted meanings. Readers are advised to regard the material as an exposition of the logical topics (a)-(c), leavened a little with names and dates for socalled human interest; and not as history of logic. At the end there is a brief section on uses and abuses of such material in the notorious history of ideas, in line with the reservations of experienced historians about this subject.

(a) As mentioned in (a) of Appendix 1, the question of *completeness* is as old as the hills or, more precisely, as modern logic. Those coming from the philosophical tradition immediately saw all sorts of difficulties, and engaged in longwinded 'analyses' of the question (in contrast to the scientific tradition, which is cavalier towards its questions, but tests its answers and their corollaries with a vengeance). How can we ask questions about the language in which we have our intellectual being as it were? Well, do we have our intellectual being in elementary logical language? Real completeness of a calculus would require that it 'embraces', that is, formalizes all valid methods of proof; cf. p. 186 (i) of [RS], with the reminder about the mind-boggling totality of all proofs of $0 = 0$. The conclusion is not to bother since, if that's 'real' completeness, it just does not lend itself to rewarding study.

Granted that the word 'completeness' is catchy, just where is the property relevant? After all, if it is considered at all, one must understand *both* the meaning intended *and* the calculus considered. In which situations is it appropriate to *combine* the two? For example, I happen to be familiar both with the model-theoretic (intended) meaning of classical predicate logic and quite a number of its formalizations. Around 1960 I asked myself: What are we left with if we forget about the latter (for the time being)? My answer was the text [KK]; cf. the long introduction to the second English or to the German translation. Other people have since tried out expositions of i.l. using primarily Kripke models, one of the concocted meanings of i.l., incidentally, not considered below. But since formal rules are mentioned, for example, in [Ga1], it is not altogether clear what we are left with if we forget them altogether in i.l. This matter is of course not settled by [KK] since the *relative* role of an intended meaning and one or other of its formalizations will not be the same in all branches of logic at all stages of their development.

(i) As long as the relevance of any — in particular, the intended — meaning is not tested, it may be scientifically frivolous to consider the completeness of a calculus for it. But for the philosophical tradition, including the variant adumbrated in section 1, it is perfectly proper to investigate the 'nature' of the matter; specifically, how much — or rather how little — need be known of that intended meaning to settle completeness; cf. Note 7a (ii) on the notion of *basis* and the historical tit bit in (a) of Appendix 1.

(ii) For classical logic, and also for i.l. (but particularly in the case of propositions about lawless sequences), I have the impression that the intended meanings have so far been almost indispensable sources of conjectures and cross checks, even when the *calculus was the primary object of study*. But it should also be remembered that interpolation and other results were first *established* proof-theoretically, and conjectured (by Craig) in connection with a socalled empiricist philosophy of science. Reminder: the latter eliminates abstract notions (from the theoretical premise) not occurring in the (empirical) consequences; cf. (c) below for more along these lines.

As to a 'correspondence' between completeness for classical and i.l., the proofs for the intended meanings came at the end of the periods considered (with incompleteness for predicate calculus of i.l., modulo Church's thesis, in the 60's). The most obvious difference was that the expositions for i.l. benefitted generally from experience with the classical case, and from such notions as basis for a more concise formulation. Another difference already mentioned is that the properties of the propositional and first order parts of i.l. are objectively more complex than in the classical case.

(b) Some *concocted meanings* of (i) classical logic in constructivist terms, and of (ii) i.l. in model-theoretic terms will now be considered. By tradition the requirement on the concocted meanings is that they should satisfy the particular laws that happen to have been formulated for the (originally) intended meanings. This tradition is dominated by such ideals as 'rational reconstruction', for example, on the ground that the original meaning is not intelligible. Those of us not limited by this particular intellectual handicap find the ideal 'unintelligible'! After all, to reword a(ii), if the intended meaning is in doubt so are the laws in question, unless they been tested in some other way. The standard claim that they represent common usage is doubly suspect. First, common usage differs from *all* formal systems, for example, w.r.t. natural sense; cf. 2d(v). Secondly, common usage is rarely obviously optimal for reasoning well.

Be that as it may, here are some items produced by that tradition.

(i) We consider first Skolem's proof of Loewenheim's theorem 'without use for the axiom of choice' (which, unlike his earlier proof referred to in Note 6a, apparently, was not properly digested by Gödel at the time, to judge by marginal notes in shorthand to that paper at the mathematical library in Vienna), and then Herbrand's *théorème fondamental;* cf. Note 7a.

Skolem's work comes under the heading 'constructivity' in the sense that attention is given to definability, in particular, arithmetic definability. So the 'concocted' meaning is that of *validity for arithmetically defined structures* (over ω). — Viewed this way there is a point to the otherwise quite silly stress on avoiding the axiom of choice; by showing that a *superstition* about the latter is involved! Specifically, the 'essential' aspect of the axiom of choice was widely seen in the fact that the other (existential) axioms of set theory implied *uniqueness* — and hence explicit definability — of the sets asserted to exist. This overlooks the fact that not even pure (classical) logic preserves this property, for example, the application of the law of excluded middle to $\exists x\,(x \in L)$: $\exists x \forall y\,(x \in L \vee y \notin L)$. Cf. also the failure of E-theorems in some 'intuitionistic' set theories.

Herbrand's *champs finis*, which are suitable sequences of expanding finite structures, are, almost as they stand, a paraphrase of logical validity, for example, in terms of an infinite Herbrand disjunction; cf. the notion of interpretation in [K1], and, in particular, of the no-counter-example-interpretation for a more compact formulation by use of function variables (avoided by Herbrand for the sake of — his particular version of — finitist ideology). Incidentally, the meaning assigned is not suitable for formulae that are not valid, in the sense that such formulae do not generally imply their own Herbrand disjunction; cf. p. 39 of [St]. — Herbrand mentions in his thesis that he knew how to deduce the ordinary completeness theorem from his analysis, and there is no reason to doubt this; cf. p. 179 of [RS]. He also added that the ordinary notion of validity was not precise enough. Here he made a philosophical mistake. The question is not 'why' Herbrand failed to prove the completeness theorem, but which kind of mistake(s) he made.

The feeble terminology 'théorème fondamental' already complained about in Note 7a, fits the feeble uses he made of the theorem; by proving some eminently forgettable prefix classes of predicate logic to be decidable. Almost 50 years later Dreben and Goldfarb [DG], quite touchingly, continued this line; though some really rewarding and of course quite different areas had been found, at least 30 years earlier, to which Herbrand's *théorème* is relevant.

If anything the point above is enhanced by the occasional success of turning the line through 180°, for example, in *un*decidability results on certain prefix classes (without any use of Herbrand's Theorem); cf. [Gol 1].

(ii) In the case of i.l., concocted meanings appeared fairly soon after Heyting's formalization, while in the classical case they had come more slowly. (Reminder: as noted in (a), completeness for the intended meaning came at the end of the period considered, as in the classical case.) The pattern is general.

Different authors were struck by — one of the relatively many — different elements of the (wild) rhetoric on i.l., cf. Note 1. Here are some samples.

As already noted in 2b there was the alleged issue between truth and provability, evidently reflected in Gödel's modal meaning 1oc.cit. Much the same applies to Tarski's *calculus of systems*, where provability (or consequence) from single axioms in classical predicate logic is meant; cf. also Gabbay's update in [Ga] in terms of Post systems.

Another catchy element of the rhetoric involved choice sequences, and such hot news as: all functions on the reals — tacitly thought of as given by choice sequences — are continuous. (Pity that Brouwer never thought of the variant: all mappings between spaces are topological!) Tarski's *topological interpretation* certainly incorporates some features of that element. More precisely, as we should say now, not so much of the choice sequences Brouwer had in mind, but of lawless sequences, which lend themselves to a smoother theory.

Perhaps the best known concocted meaning is recursive realizability and its variants; in the first place with stress on the constructivity of operations, not of proofs. Kleene was (consciously) most taken by a felicitous formulation of Hilbert about implication involving partial information. As far as content is concerned, this is barely distinguishable from formulations by Brouwer and Heyting (and not even appropriate for the finitist case, Hilbert's avowed concern, where the formulae A, B are Π^0_1 and $A \rightarrow B$ is realized by a total recursive functions). But Hilbert's wording fits the need, already mentioned in the *Digression* for specialists in 4a (iii), to use partial recursive functions for realizing the logical laws.

Stretching matters just a shade perhaps, one might look at the enrichment of realizations by formal derivability in Kleene's q-realizability, and especially by formal derivations in Beeson's fp-realizability some 30 years later. Both reflect the concern of — some of — the rhetoric with proofs over and above definitions; specifically, proofs of the fact that the definitions do what they are supposed to do.

In summary, here we have examples of exploiting an intended meaning without unduly meticulous attention to it. This style is clearly appropriate if one is convinced that the details of the intended meaning do not serve the intended purposes or that the latter are misguided.

(c) The topic of the *relevance of intended and discovered meanings* was already touched above, in A2a (i). Here are some reminders and examples.

First of all, the most commonplace activity in mathematics, generalization, is a discovery of a new meaning (among other things). $a^2 — b^2 = (a + b)\,(a — b)$ may originally have been intended for a^2 and b^2 ranging over ω, but is discovered to be valid for all commutative rings.

Secondly, the philosophical tradition of working up dramatic conflicts or at least puzzles — here, concerning the meanings mentioned — is quite misguided. (This is of course a recurrent theme of the whole article.) There is no general conflict, but there is a very real problem: Where is which of these meanings, if any, relevant? There is no puzzle, for example, about different meanings being equivalent (in the here usual sense of having the same set of valid sentences in the languages considered) since those languages obviously do not exhaust the possibilities of talking about those meanings. The tradition is also misguided in a more practical sense since those antics about conflicts and puzzles attract philosophical cripples, for example, among — both gifted and other — logicians; 'misguided' unless the tradition simply wants to preserve the status quo.

It is equally misguided to dub the choice between those equivalent meanings a mere matter of convenience, compared to the allegedly primary matter of adequacy-in principle. This is generally simply not good enough for *effective knowledge;* cf. (i)-(iii) below or consider the choice between walking and taking a car to, say, the nearest source of food; if the latter is 1 km away the choice may be a matter of convenience, if 100 it is a matter of survival. So much for adequacy-in-principle.

The next items, though still general, concern the intended meanings of classical and i.l. Quite simply, the usual logical languages express very little about those meanings. Thus nothing about the power set construction even enters into the explanation of the (classical) meaning of first order formulae although that construction with its iteration may fairly be said to be the heart of the subject of sets, in terms of which the set-theoretic, alias model-theoretic meaning is defined. At best, knowledge of that construction may be used in *logically impure proofs* of logical validity.

In the case of i.l. the intended meaning involves constructions with their (decidable) properties and proofs of assertions that any — that is, an arbitrary — construction has some given property. But for logic the principal topic is the relation between a proof and the assertion proved. Realistically speaking, this relation is just too meagre to be rewarding (or even to save one from oversights like paradoxes because it just does not provide enough cross checks.) Taken literally, it leaves no room at all for more delicate ideas about proofs such as — those on p. 187 of [RS], nor even for — Brouwer's about 'fully analysed' proofs in 4e (i) (and their extension in Gentzen's proofs without detour). If these ideas are to be pursued in a purely logical context at all, some concocted meaning is *necessary,* for example, socalled operational semantics. And then at least the scientifically experienced will ask *whether* the most fruitful aspects of those ideas are relevant to logical contexts at all. It would be pure ideology to assume this just because the word 'logic' is glamorous.

To conclude, here are some specific examples, chosen at random, that are related to the generalities above.

(i) Elements from the constructivist tradition, especially, the part concerned with definability are practically unavoidable even in classical logic as soon as the languages considered are restricted, for example, to finite formulae. As already mentioned on top of p. 179 of [RS], Herbrand's pre-occupations described in b (i) above have turned out to be useful even when applied to Σ_1 formulae; admittedly, once stated, the form of a Herbrand disjunction for $\exists xA$,

$$A\ [x/t_1] \vee \ldots \vee A\ [x/t_n]$$

is memorable without (the no-counter-example-) interpretation, let alone, the more tortuous *champs finis*. — Update: since [RS] appeared, applications of those ideas to Σ_2 formulae have been found where already the wording of the Herbrand disjunctions benefits from functional interpretations. — *Reminder* for specialists. Once the wording is available, model theory is quite adequate to infer the *validity* of some Herbrand disjunction from the validity of the formula in question. Herbrand himself aimed at — but did not achieve himself — *quantitative estimates* for a suitable disjunction from richer data, namely the validity of the formula enriched by a suitable proof.

(ii) It is a common place that many formal results about i.l. have been obtained by use of concocted meanings; cf. b (ii) above. But it is worth stressing that also the (first) completeness proofs for the *intended* meaning, mentioned in A 2a (ii), used the topological interpretation concocted by Tarski; cf. also the exposition by Burgess [Bu] using a meaning concocted by Kripke. The proof for the intended meaning required 2 further steps. First, the observation that, instead of 'associating' an open *subset* A_p of a certain topological space S with the propositional symbol p, the latter can be literally interpreted, that is, replaced by a *proposition* A_p with parameter α over S. Secondly, if S is chosen as a (topological) space of lawless sequences α then Tarski's 'association' is proved to respect the laws of lawless sequences for propositional operators o: $\alpha \in A_{poq} \Leftrightarrow (\alpha \in A_p)$ o $(\alpha \in A_q)$. In short, far from having a 'conflict' between the concocted and intended meanings, one has a *proof of equivalence* in the relevant context.

(iii) This is a speculation on a possible use of the many concocted meanings for formal i.l. (in accordance with the rhetoric about formal laws not determining meaning, but at odds with the wailing about this truism). It is addressed to those who wonder whether $P = NP$, and believe that the question is 'logical'; for example, as opposed to: belonging to transcendence theory; here one would look for some familiar real number with, say, a decimal shown to be both *NP* complete and in *P* or not (as the case may be). Now, with each concocted meaning of i.l. comes a whole body of knowledge: of the concepts used to define that meaning. So there is a chance of finding a problem about i.l. that can be proved *NP* complete from knowledge of one concocted meaning, and decided to be polynomial (or not) from knowledge of another; cf. the *Remark* at the end of 3b about equivalent definitions of 'recursive'. NB. *Not the similarities* between different meanings, so dear to logicians' hearts, *are useful* here, *but differences*.

(d) It would be unrealistic to rely on the disclaimer at the beginning of this appendix as a safeguard against misinterpretations of (a)-(c). On the one hand there are stories about individuals, often chronologically ordered; the sort of thing associated with history. On the other hand there are subdivisions, and subdivisions within them, associated with socalled scholarly history as opposed to, say, historical novels or to pep talks using historical illustrations (practised by politicians in all spheres of life). In the long run, any attempt to use the snippets in this appendix or, for that matter, in the Notes as a conventional kind of history would draw attention away from the more modest, but surely worthwile uses actually made of them here. The reason is by no means marginal. It will now be considered at leisure.

Without exaggeration, our ordinary view of men and their doings is extremely primitive. Sure, all the world complains about lack of 'progress' in our understanding of human nature and human society, compared to progress in the natural sciences. Some intrinsic obstacles are similar to those already mentioned at the end of 2d (vi) in connection with the uphill fight of natural history. But the overwhelming obstacle is probably the obstinacy with which one clings to those aspects that strike our untutored attention; in fact, literally to the *Wisdom of the Ancients* in Note 3a; here, to the Bible with its list of sins, another word for: principal human motives. If somebody comes along and selects one sin as *the* dominant force this is regarded as an intellectual revolution; one used to think of Marx and Freud, but one might as well throw in the analogue to purity of method already cited in 4f (i). (Of course, genuine revolutions came about when this kind of rehash was presented with uncommon conviction in uncommon circumstances, and, perhaps above all, with uncommon political, including literary skill.) Just imagine one had made a virtue of clinging to ideas of the same vintage about the world around us! Surely, there is room for a dialogue in which honest *Simplicio* tells us how he looks at human phenomena, even if he only says what everybody (still) says.

Now, of course we live with this sort of thing, having intellectual and other reflexes that operate independently of traditional commentaries. What the socalled history of ideas adds to this mess is the pretense that *painstaking scholarly documentation* — comparable to the most simple-minded painstaking classifications in natural history, cf. 2d (vi) — improves matters. On the contrary, one adds to the simple-minded view a crass imbalance between the degree of accuracy implicit in the claims and that of the data; cf. Note 8 about documents related to Gödel himself (but also about the possibilities of settling convincingly imaginatively chosen historical questions).

Superficially, all this applies to all history. Not so; gifted historians have discovered — either, again, by exercise of sound intellectual reflexes or, occasionally, by following a 'philosophy' of history — aspects that do lend themselves to rewarding study; most spectacularly perhaps in archaeology, obviously related to history (even if separated, partly for ideological reasons); in particular there are uses of archaeology in connection with Babylonian mathematics.

In accordance with common experience — so to speak, as a lesson of history — a subject like the history of ideas will be populated by people who are prepared to make little progress provided only few others make more; in particular those one-eyeds who have such a good time in the kingdom of the blind. Besides, it is a *pleasant* subject. Even if no new knowledge results, at least, one is paid for reading material by gifted people; comparable to natural historians who spend their time looking at pretty butterflies or flowers instead of bacteria looked at in scientific genetics.

Far be it from me to claim that molecular biology is the sole key to all biological, including human phenomena. But I at least am grateful to have lived through a period when the conventional threadbare literature on human nature in socalled social sciences was replaced by a fresh kind of speculation, even if the latter is often only a *jeu d'esprit*; cf. p. 218 of [RS] for a new twist on *Genesis* by Crick and Orgel. It concerns history; but not the history of ideas.

To conclude, it is salutary to remember one of the great successes of the natural sciences in the area of historical research. Here I shall take the history of the planets since most readers will already have thought of the — origin and — history of species on earth. Naturally, *Simplicio's* future dialogue, mentioned above, would tell us that planets are not conscious beings who intend to leave a record of their doings (and he will feel that he has made a profound and, above all, decisive point). It never occurs to him even to try and *test* how much help this is; cf. Note 5d *in fine*, but also Note 7a on what Herbrand and Gentzen knew, but could not say.

There are two aspects to the history of the planets.

(i) The history of their *outward* behaviour, in other words, their motion. This admits, as one says, astronomical precision. *Reminder*: Not their apparent motion, so to speak, what they say literally, but the motion corrected for parallax (by Tycho Brahe) and so forth; cf. the *Digression* in 2d (ii).

(ii) The history of their *inner* life, in particular, their chemical evolution. This not only can be, but has been asked and speculated about for a very long time. But it is fair to say that without a rather advanced knowledge of subatomic processes one had not even reached the kind of *threshold* that came up in Note 3b (i) and several times since in this article.

Remarks. The other great success, the history of species on earth, seems — as far as I am competent to judge — more delicate. No doubt some evolutionary phenomena, the spread of epidemics or development of resistance to drugs in bacteria, lend themselves to a theoretical study comparable to (i), that is, using little more than Mendel's laws and more or less sophisticated mathematics (as in particle, resp. continuum mechanics). But knowledge comparable to (ii) — on the molecular if not subatomic level — is liable to be relevant to the evolution of outward forms, even if only quite rough approximations are wanted. For example, just how many one-step mutations are needed for some morphological change? The answer would seem essential for giving an even approximately sensible meaning to 'missing link'. — Readers are asked to forgive the traces of 'metaphysical anger', in the sense of Note 1b (ii), that they may find disturbing in this section (d). But what more do those glib would-be historians (here, of logic) think they know of the relevant intellectual mechanisms than about mutations?

NOTES

1. *Intuitionistic rhetoric*: some impressions and reminiscences. The presentation in section 1 of i.l. — as a modest correction of syntactic, that is, purely external defects in the logistic scheme — may be sounder than the popular intuitionistic rhetoric, but it is less memorable. The orthodox rhetoric associates the difference between classical and i.l. with general philosophical 'positions' on the nature of mathematics (or of the world itself, if you want to go the whole hog). The key words are objective and subjective. They are familiar, in fact, hackneyed, and hence all the more memorable (even if one does not know too well what one remembers).

(a) Reactions to the rhetoric vary. One extreme is Bourbaki's, reflecting undoubtedly the view of the silent majority: i.l. is an historical curiosity, tacitly, to be ignored. (It is a view of the rhetoric, and not of the details of i.l. because these never got known.) — An opposite extreme is the dramatic thrill of a conflict, even though already Georg Lichtenberg put the orthodox 'issue' above in perspective in his Aphorismen.[6] But also the deeper thrill of

[6] ...Kantische Philosophie ist die gewiß wahre Betrachtung, daß wir ja auch so gut etwas sind als die Gegenstände außer uns. Wenn also etwas auf uns wirkt, so hängt die Wirkung nicht allein von den wirkenden Dingen, sondern auch von dem ab, auf welches gewirkt wird... Kantischer Geist... die Verhältnisse unseres Wesens... gegen die Dinge [,die wir] außer uns [nennen,] ausfindig zu machen; das heißt, die Verhältnisse des Subjektiven gegen das Objektive zu bestimmen. Dieses ist freilich immer der Zweck aller gründlichen Naturforscher gewesen, aber die Frage ist, ob sie es je so wahrhaft philosophisch angefangen haben wie Herr Kant. (If the parts in square brackets are kept one merely has relations between different parts of socalled subjective experience; not an equally clean separation. Whatever else 'wahrhaft philosophisch' may mean, it is pretty certain that no natural scientist has succeeded in spreading out Kant's *Betrachtung* over 700 pages. In particular, nobody before him has succeeded in giving comparable weight to Kant's reminder, a matter not to be despised.)

indignation is to be remembered here, which was triggered by the 'menace' of i.l. At another opposite extreme there are studies that can be seen as investigating the rhetoric, even if the authors had different principal or at least additional interests; cf. (b) and (c) below. For example, (i) for philosophy in the sense of section 1, one may wish to put the rhetoric in its place or (ii) one may wish to see if anything of any sober interest can be extracted from the rhetoric at all as, for example, I tried from 1958-1963 (or even something wonderful, as vendors of sheaf models seem to suggest).

(b) Whatever Gödel's research interests may have been, the styles of his early and later presentations differ sharply, as already mentioned at the beginning of this article. As elaborated in section 2 the early notes are concise and cavalier, apparently scoffing — by contrast, cf. also p. 154 of [RS] — at the antics of the rhetoric. Later on, even where he disagreed, his comments on any kind of traditional philosophical concerns were respectful to the point of reverence; cf. bottom of p. 172 of [RS]. Here are a few possibly relevant odds and ends:

(i) Though later — cf. top of p. 174 or p. 197 of [RS] — Gödel used crude, hackneyed formulations that had proved to have popular appeal (and had put me off, cf. p. 158 of [RS]), in his very early writings he was more austere. In the introduction to his dissertation he scoffed at the *Grundlagenstreit* (which Einstein had called a cat-and-mouse game), and soon aftetwards, at Königsberg, he treated (Hilbert's) claims that consistency was a sufficient condition for soundness similarly; cf. bottom of p. 153 of [RS]. — Also, he was offended by thoughtlessness; at least, this is the way I interpret the following item in Olga Taussky's piece in this volume.

(ii) Even after Gödel's incompleteness paper Hilbert continued to repeat the mantra:

(*) it is consistent to assume that every problem is solvable,

in other words, that for every proposition *P*, either *P* is provable or–*P*.

Now Gödel's second theorem shows even more:

(**) it is consistent to assume that every proposition is provable (in the system considered). And Hilbert never saw that (**) $\Rightarrow$ (*)! This is not merely a personal, but a metaphysical affront!

What is so offensive here is that Hilbert had been repeating his 'grand' pronouncement with the conviction that (*) was so deep, it could never be refuted; obviously never dreaming that it could actually be proved! and trivially to boot. (And it would be absurd to split hairs by saying that (*) is not proved by the means of proof for *P* considered since Hilbert's many false conjectures are quite implausible once the relevant difference in methods of proof is remembered at all.)

This is the kind of stuff of which metaphysical anger is made.

(iii) Since his early student days Gödel certainly pursued theological and other broad philosophical topics. But I myself have not seen any record of a similarly early interest in ordinary academic philosophy, for example, in foundational debates related to the Grundlagenstreit mentioned above. On the contrary, in the introduction to his dissertation he used the notion of validity as a matter of course, and later truth of arithmetic statements. This is very different from a non-constructivist position, which makes an issue out of accepting those notions; as indeed Gödel himself did in his later popular writings noted in (i) above; cf. also Note 5b on other instances of Gödel's style in conflict with the academic philosophical tradition. — For the record it may be mentioned that, when Gödel said the word 'philosophy' with a trace of awe in his voice, his wife reminded him of his habit, back in Vienna, of stressing that he was a mathematician (incidentally, imitating his voice quite successfully). Actually, this habit would have served a good purpose of keeping philosophical pests at a distance; cf. p. 159 of [RS]. Appendix 1 contains more about Gödel's 'non-constructivism'.

Towards the end of his life he is quoted to have said that i.l. was bad for mathematics, but important for foundations. (Before 1970 he never made such unbusinesslike remarks to me, at least, not in logic; cf. p. 160 of [RS].) The following twist is in line with this article. I.l. has not so far proved to be a useful tool in the arsenal of mathematics, though it has been a quite rewarding object of (meta)mathematical analysis; cf. Note 2. On the other hand it is a gold mine for foundations in the sense of section 1, that is, for examining extreme ideologies. NB. This was not Gödel's sense, above all, not towards the end of his life when he confided to me that he expected foundations (and philosophy generally) to tell us what the world is really like; as if science did not have this goal, too.

(iv) Finally, for the record as it were (not to impose some artificial order of cause and effect), here are some personal matters concerning Gödel's reactions to Brouwer's style.

Menger, who strongly encouraged Gödel from the start, had spent some time in Amsterdam where he had become sensitive to some weaknesses of i.l.; possibly helped by a long forgotten priority dispute with Brouwer about the (obsolete) business of a correct definition of dimension, that is, Menger was helped to see the splinter... as He said; cf. the review of the whole journal *Fortschritte der Mathematik* on pp. 6-8 in vol. 39 (1932) of the *Mhte. Math. Physik.* Once again, be that as it may, Gödel's elegant formal results, discussed further in 2a, could be interpreted as making Menger's impressions more precise. It would be entirely in keeping with my own experience of Gödel as a staunch friend — when I disturbed some people in the 50's by what were then considered to be indiscretions — if he had taken this opportunity to support Menger. After all, the results he used continue to be quoted.

At least in the 50's Brouwer's personal style, of haranguing his audience for hours, did not suit Gödel at all. Gödel complained about having to play the host; also, if I remember correctly, in one of his letters to his mother. Gödel was utterly bored by Brouwer unlike several logicians and mathematicians who, being dry themselves, were buoyed by Brouwer's probably genuine exuberance; cf. bottom of p. 158 in [RS] on Gödel's reaction to exuberance. — I never asked if he attended Brouwer's lectures at Vienna at the end of the twenties (as he presumably did). If he did, his reaction would certainly not have been very different. Gödel, incidentally like Brouwer himself, did not change his tastes, and was proud of it; he called any change of taste: *Mangel an beständigen Gefühlen.*

I do not suggest that these personal matters could be decisive. But if one's confidence in an enterprise like i.l. is shaky to start with, the performance of its chief exponent can give one the final push. I have a relevant anecdote.

(c) My first encounters with Brouwer's style were in the late forties, at the first of his lectures at Cambridge, and then at his (invited) lectures at

University College in London. I was utterly bored by his exaggerations, and asked him after a lecture if he meant all he said. He quoted George Bernard Shaw on having to exaggerate to make an impression, in a style that made me feel he had used the quotation repeatedly; cf. b(ii) on Hilbert's repetitions. I pointed out, as innocently as I could, that G.B.S. had not promised him that he'd make a good impression. Incidentally, Brouwer was not amused. Apparently, he did not like to be interrupted anyway; fittingly, for a good solipsist.

Now, I certainly was sceptical of i.l. before I ever met Brouwer. For example, I am on record stressing the appeal of the more radical restriction to quantifier-free, in particular, finitist schemes if and when it is appropriate to be constructive in mathematics at all. And I had never taken seriously the principal preoccupation of foundations alluded to already in section 1, of exhibiting the logical laws implicit in ordinary reasoning. (Philosophers, who have this preoccupation and are interested in constructive aspects, feel obliged to be interested in i.l. simply because logical words occur in that reasoning.)

Sure, there wasn't much for Brouwer to spoil in my case. But I do remember that a phrase I used quite often in later writings, occurred to me during one of Brouwer's lectures: those iterated implications make my head spin; as they still do; just like higher types, their counterpart in logic-free mathematics. — *Reminder*: the formal theory of those things, even in ramified set theory, is quite elegant (as long as one does not think of instances).

2. *Disjunction and existence properties*. Background. By now it is fairly generally recognized that these properties are not — or, at least, not generally — required by intuitionistic validity. On the formal side there are systems that do not have them; some were manufactured for the purpose; some introduced for other purposes were discovered not to have them. More instructive are the following reminders. (i) In the case of logic, where propositional and predicate symbols are interpreted as variables (when a formula is called valid), $A \vee B$ is prima facie comparable to $a = 0 \vee a \neq 0$ in number theory. The latter is true for all a, but neither $\forall a\,(a = 0)$ nor $\forall a\,(a \neq 0)$. The obvious difference is that, in the case of general validity — tacitly, for a wild stock of propositions or of domains and predicates in predicate logic — too little is known about these things to use their structure

in associating effectively either A or B with each value of the variables. In that case, either A can be asserted outright or B; cf. the impossibility of separating the continuum 'continuously' where the characteristic property of continuity is that only very limited information about arguments is used. (ii) For (arithmetic) formal systems with a specific interpretation, incompleteness intervenes. So if a closed formula $A \vee B$ is derivable and the system is sound either A holds or B; but, by incompleteness this does not ensure that either A is derivable or B. So if a system has the disjunction property its incompleteness w.r.t. disjunctions balances as it were its incompleteness w.r.t. the disjuncts. — The following remarks concern Gödel himself.

(a) He probably noticed quite early the facts discussed in the last paragraph, but I am not sure. As it happened, I noticed them before I met him, and mentioned them to him soon afterwards. Now, he always took pleasure when somebody else spotted a point that he liked himself, and — at least, in my experience — never mentioned his independent discovery, let alone, priority (even when the matter was in print so that it was irritating for the other not to have the reference). For example, he knew very well my aversion to Hilbert's claims for consistency; incidentally, for exactly the same reasons that he had given at Königsberg at a time when I could barely spell 'Widerspruchsfreiheit'; cf. Note 1b (i). But he never referred to that lecture, which I read for the first time when preparing [RS]. — The next anecdote will be elaborated in section 4.

(b) By the end of the 30's Gödel had doubts not only about the existence property of Heyting's formal arithmetic HA, but this: Does a formal derivation d in HA of $\exists x A$ ensure some term t_d, defining a number x_d, such that x_d satisfies A (without $A\ [x/t_d]$ being necessarily derivable in HA). This was the main problem that the material in the 40's that later developed into the Dialectica interpretation was said to solve (in the relevant notes for a lecture at Yale in the Nachlaß). Two points about this will be used below. (i) At least prima facie Gödel did not solve this problem for the interpretation intended by Brouwer and Heyting. The term t_d ensured that $A\ [x/t_d]$ was valid for the Dialectica interpretation; in particular, it was not at all apparent that $A\ [x/t_d]$ was formally derivable in HA. (All this and more was settled in the 70's.) — (ii) On several occasions Kleene has coyly referred to a 'well known logician', evidently meaning Gödel, and his doubts about the disjunction and existence properties. But Kleene never elaborated just what was being doubted; neither the distinctions made in the first paragraph of this note nor the point in (i) above.

(c) This reminiscence concerns the unhappy seventies, cf. top of p. 160 in [RS]. H. Friedman had shown that all formal extensions of HA with the numerical disjunction property, also have the existence property; the converse being obvious since $(A \vee B) \leftrightarrow \exists x\,[(x = 0 \rightarrow A) \wedge (x \neq 0 \rightarrow B)]$. Gödel eventually submitted the paper for publication, but only after worrying whether the result was really completely general. He had to be reminded of the (good) reasons for his worry! They go back to the widespread belief that the properties in question are needed for intuitionistic validity. Given this blindspot there was suspicion that some tacit assumption had slipped into Friedman's proof and restricted the systems from the start, thus trivializing the result. In fact, when the paper was published several outsiders were ill at ease about the paper just because of the blindspot. The paper is of particular interest precisely because those properties are *not* needed for intuitionistic validity. In short, the paper is not a mere curiosity, contrary to the impression conveyed by [NH]; cf. also Note 7c (i).

3. *Unpopular traditions,* but also: marginal elements of the scientific tradition, the latter being by no means homogeneous. There is one element in science, repeatedly cited in this article, that really strikes the outsider. It is the selection of, often not all appealing aspects of phenomena. In fact some of the gushing talk about 'freedom of science' fits here, if interpreted as freedom to try out different aspects to find those (few) that lend themselves to extended or even theoretical study. Incidentally, this activity would have its counterpart even in a cruel free enterprise style of science where scientists are rewarded like investors or enterpreneurs, and not simply paid to do what they like or coddled in other ways; for the view of Gödel's wife on coddling, cf. bottom of p. 154 of [RS].

The unpopular traditions in (a) and (b) below, under the titles 'wisdom of the ancients' and 'natural history' tell us not to work too hard at those selections. Instead one is to rely, as the name suggests, on venerable traditions, resp. on immediate appeal; either as explained at length in 2d (vi) or on appeal to the senses. Gödel was particularly fond of the former. We begin with an anecdote.

(a) Wisdom of the ancients: including an interest in ghosts, demons, deities and so forth. When I read a footnote on time travel in Gödel's article in the Einstein volume it struck me as attempting to provide a kind of explanation why one rarely sees familiar ghosts, that is, of the recent past; cf. p. 215 of [RS]. I happened to tell him about it in the presence of his wife, who then spoke mockingly and at length about his life long interest in ghosts (which, according to her, was shared by Viennese washerwomen), and about the many books he had read on the subject. (I remember using this to change the subject by noting that those washerwomen surely did not rely on books.) Gödel became very expansive on the need for a great deal of basic agreement if conversations are to be fruitful; but cf. (iii) below. I did not need much persuading since I have always applied this 'non-missionary' view even to my writings. Here are a few additional points.

(i) Gödel's faith in the wisdom of the ancients was not shocking to me. It reminded me of a lecture by Lord Keynes at Trinity College, Cambridge; actually, the first — and almost last — evening lecture I ever attended. Keynes had bought in Ireland a ship trunk full of Newton's papers, and spoke most memorably of Newton's attempts to deduce all sorts of things from the number of the beast (in the Apocrypha). I tried to convey my impressions of Keynes' style to Gödel, but doubt that I succeeded.

(ii) Without sharing Gödel's faith at all I did not reject it out of hand; at least, not abstractly. Obviously, unusual skills were needed to make good use of it. Common place objections to that kind of faith seemed to me weaker than the faith itself; perhaps, comparable to my distaste for the usual objections to informal rigour in the analysis of intended meanings, even granted that the latter may be unsuitable for their intended purposes. — Besides, as elaborated in A2d, in human affairs the wisdom of the ancients is widely accepted; in effect, if not in these words.

(iii) But the agreement in (i) did not go very deep. On one occasion, I think out of the blue, Gödel brought up the familiar asymmetry of the universe: so many more events are unpleasant. From this he concluded the existence of demons. I don't remember, if ever I knew, what came over me to talk about an evenly distributed universe, but my being so often below par that I could not exploit pleasant opportunities. Evidently I had forgotten to consider the role of demons in my being below par. At any rate, in his gentle way already noted on p. 155 of [RS], Gödel attributed my blunder to lack of interest, not demons. We never talked about the subject again.

(iv) There clearly was something to Gödel's view, at least, if the loaded expression 'wisdom of the ancients' is replaced by, say, 'naive ideas'. Gödel's own favourite refrain was: if one could be so successful [with such ideas] as he was, one must expect marvels if one tried harder. During his life I never felt quite comfortable about the whole business. Only when I came to the end of [RS] did I put my reservations into words. Perhaps those ideas are good to remove blindspots, and then they are wonderfully efficient. But it still bothers me that the law of diminishing returns seems to apply to them so very soon.

(b) Natural history; cf. also 2d(vi). This subject came up only rarely, and then tangentially; in accordance with general experience that one minority sect — here, those faithful to the wisdom of the ancients — tends to be suspicious of any other. The following reminiscences involve natural and computer languages.

(i) I myself have always had a soft spot for Chomsky's and René Thom's styles; the former simply by contrast with his critics, the latter because of his aperçus. For example, the matter of natural sense in 2d (v) was either mentioned by Thom or occurred to me during one of our conversations (in line with — not in opposition to — his thought, in contrast to the end of Note 1c). Gödel was sceptical.

He was disturbed by lack of precision in Thom's writing, of which he had seen only a paper on language. He once spoke about a very long typescript of Chomsky's that circulated in the sixties and consisted almost wholly of critical observations. Though formulated in quite different terms Gödel's reservation was that the subject simply had not reached the *threshold* where this kind of detail was rewarding. Without exaggeration, much the same applies to the bulk of natural history, which of course is particularly proud of its painstaking detail.

(ii) On a couple of occasions he mentioned computer languages, presumably after the subject had come up in conversations with others. It has been reported, for example, by Zemanek that Gödel more or less advocated predicate calculus as a programming language. He never suggested anything like that to me. (When bored Gödel often made offhand suggestions, for example, conjectures to visiting members at the Institute of Advanced Study, who would then spend hours doubting their own, often completely trivial proofs or refutations of those 'conjectures'.) But he did say — of course, expressing a mere feeling, without any basis in experience of the subject — that programming was *Sache der Geschicklichkeit*, in other words, a skill, and not likely to benefit from theory at all, let alone, logical theory. The word 'skill' jars since practically everything needs some skill in his sense; cf. the digression at the end of 2d(ii).

Actually, one can be more specific here, by reference to *Prolog*, short for: programming in logic. It is successful; not because it uses predicate logic, but because it does *not use all* of it. This is verified by studies of various attempts to add negation (to the Horn sentences used).

Be that as it may Gödel certainly did not expect programming to benefit from theoretical studies of natural languages; or, more pedantically, from realistic theories. A bad theory — so to speak, how der kleine Moritz imagines natural languages to function — may well contain a bright idea that has some use for some program for some hardware for some computational problem. (As somebody once said in a paper on the nervous system, with a far-fetched theory of r.e.m. dreams: if Nature does not use our idea, perhaps it can be used somewhere in A.I.)

4. *Effective rules:* supplements to section 3 and earlier literature. The topic came up repeatedly in conversations with Gödel during the sixties, but was not pursued. Whatever the reasons may be, our interests were certainly very different; recall Note 3a. He was preoccupied with giving a conclusive proof of the difference between — effectiveness for — mind and matter; cf. p. 216 of [RS]. My principal (conscious) interest was to give vent to the kind of metaphysical anger mentioned in Note 1b (i). Specifically, both extremes — in current jargon, wide-eyed enthusiasts of A.I. or, better, of digital intelligence, and indignant critics — never give a thought to the possibility that *we know enough already to refute their rhetoric;* but, presumably, not enough to settle any significant issue. (In other words, we may not have reached the kind of *threshold,* for genuine progress, mentioned at the end of Note 3b (i).) — We begin with a supplement to p. 216 of [RS].

(a) During a stay at the Institute for Advanced Study in the sixties I prepared the survey article [Sa] (about which I spoke very little to Gödel since he regarded such activities as a waste of time). Footnote 29 on p. 144 mentions the possibility that, at least, statistical mechanics may be demonstrably non-mechanical; cf. also bottom of p. 2 16 in [RS]. I returned to the topic — of rules or, equivalently, inputs that are effective for physical systems for which a theory is available — in [K6] and [K10]. The former excludes some plausible candidates for systems that have non-recursive outputs for some recursive inputs, by adding physically relevant conditions to a simple-minded formulation, specifically, in *percolation problems*. Reminder: a Markov process with unique asymptotic behaviour, the additional requirement being *non-vanishing* probability of that behaviour.

Since then also physicists have shown an interest, for example, Geroch and Hartle in the Festschrift for Wheeler. The logically more conscientious work on the topic above, by Pour El and Richards, is however spoilt by lack of respect for the physical meaning; cf. the review of [PR] on their abuse of the notion 'initial value'. On the positive side their work has improved my understanding in 2 ways.

(i) On the formal side the boring lists at the beginning of the review of [PR] — of recursive operators that do and do not necessarily have recursive values for recursive arguments — can be subsumed compactly in terms of boundedness; cf. [PR1], and also the more colourful terminology of (Gödel's) growth in 3b(ii), and Turing's 'philosophical error' in c(ii) below.

(ii) More generally, [PR] refutes the idea — I had for some time (without mentioning it in print) — that people working on a problem, say, the 3 body problem, might make a tacit assumption: sound solutions *must* be recursive in — what are regarded as — the initial data. In this way they would simply miss non-recursive solutions. [PR] shows that this was not the case with Kirchhoff's solution of the wave equation. — NB. Non-recursive solutions are often indeed unsatisfactory as they stand. But once recognized they may be explicitly excluded for physical reasons (parallel to the use of Gödel's solution of Einstein's equations according to the letter of Penrose quoted on p. 215 of [RS]) or they may suggest *new questions* that have more manageable solutions; cf. the famous example in number-theory where there was only the *suspicion* that no simple method decides whether a binary diophantine equation has *some* integral solution. So Siegel asked instead if it has *infinitely many,* and decided that question.

(b) Before examining the particular rules for the perfect (intuitionistic) mathematician described in 3c (i), I looked at another area of formal work with the potential of being relevant to the matter at issue: the

consistency of — suitable formalizations of — Church's thesis (CT) with intuitionistically interpreted systems, like the theory of species.

As might be expected this was more to Gödel's taste than — what he, by Note 4a (ii), would have considered as — *Kleinarbeit* on the stability of *E*-theorems in 3c (i); cf. (iv) below. Actually, even an inconsistency would only show that CT cannot be *proved* by methods in the system considered. It would not necessarily furnish a rule that is effective for the perfect mathematician, but defines a non-recursive function. — Before I dropped this subject I summarized my experience in [K 9]. Here are a few points from [K 9], and an open question not considered there.

(i) If a system has the *E*-property, and is *formal,* that is, the set of theorems is r.e., then for any theorem $\forall x \exists y R$ there is a recursive function f s.t., for each $n \in \omega$, $R(\bar{n}, \overline{f n})$, is derivable.

So, if only formal systems are regarded as 'precise' the *E*-property excludes the most direct refutation of Church's thesis indicated above; that is, by a proof of $\forall x \exists y R$ that is convincing for the perfect mathematician, while no recursive f satisfies $R(\bar{n}, \overline{f n})$ for all $n \in \omega$. — Obviously, it would be a *petitio principii,* in connection with effective rules, to *assume* that only formal theories are precise.

(ii) The idea of a creative subject proceeding in an ω-series of steps, and the 'axioms' for the property $\vdash_n A$ (A is proved at stage n) implying Kripke's principle are found to be implausible. Roughly speaking, ω is too easily grasped.

(iii) An opposite extreme as it were is the idea of *transfinite progressions*, originally called 'ordinal logic' by Turing. In fact, the title of [K 9], which refers to this topic, is related to conversations with Gödel that are not mentioned in the paper.

In the early seventies, at a very difficult time in his life, he made several very long, mildly frantic transcontinental telephone calls (from Princeton to Los Altos Hills, where I lived at the time) about the present topic. He felt sure that, for *every* path through O, *some* non-recursive predicate is computable (on any transfinite progression of the kind studied by Feferman). Gödel *regarded this as a refutation of* CT. It was not necessary to go into his idea since it was practically apparent from the paper by Feferman and Spector that Gödel was simply wrong:

No non-recursive predicate can be computed on any Π_1^1-path.

He insisted that this should be published, and I simply did not have the heart to refuse. He agreed that, since the proof was so close to published material, and the result so inconclusive it would be out of place to describe it as a refutation of a conjecture of his. I had done such a thing in [K 4], partly through carelessness, partly for reasons explained at the beginning of A1 (c); actually Gödel's offhand habit, alluded to in Note 3c (ii), was only brought home to me a few years later by reports of a victim who spent a year at the IAS.

My compromise, some 15 years later, was to do what he wanted in (the short) Part I of [K 9], with an appropriate title, but use Part II for what I wanted to do myself.

(iv) Here is the still open question alluded to at the beginning of (this) Note 4b; recall also 4e (i) of the text. It concerns the *consistency of Church's Thesis* with socalled *extended* bar induction EBI, which first arose in the formulation of bar induction of finite type applied to — neighbourhood functions of — continuous functionals; cf. 4e (i). The question is, incidentally, the only problem on pp. 156-157 of [K 5] that is still open. However, an up-date is needed.

The original question concerned EBI added to the theory CS for choice sequences with *analytic* data. But there is nothing sacrosanct about those data; after all, they were preceded by *open* data. On the contrary, it is part of the question to *select data* for which the matter of consistency with both EBI and CT is rewarding!

The few contributions on EBI have been inconclusive (and of quite uneven quality). A memorable point, noted in [T 1], is that, for suitable data, the premise $(\forall\alpha \restriction a)\, C^{\sigma}(\alpha)$ in 4e(i) is *almost negative;* for many (realizability) interpretations this ensures 'independence of premises', and is thus special among the general cases of EBI. But there are defects in [T 1], apparently discovered by Renardel (though not listed in his dissertation [Re]). On the other hand, [Re] confines itself to such a special case of EBI that it has the same arithmetic consequences as BI! In other words, this case does not even touch the most highly advertized difference between BI and EBI (applied to the neighbourhood functions mentioned above), namely, the much greater proof-theoretic strength of the latter.

(c) Both [K 9], pp. 318-319 and Gödel's posthumously published remarks on the undecidability results mention Turing's errors or, better, tacit assumptions; cf. p. 250 of [Tu]. Of course, they are trivial compared to Turing's contribution: his focus on rules effective for computing *machines* had raised the level of the discussion not only far above the drivel about equivalent definitions, but also above Gödel's discovery of absoluteness; cf. 3b of this text. Here are a few additional details.

(i) Gödel had surely noticed the *petitio principii* mentioned in b (i) — of assuming that only formal rules are precise — that struck me so forcibly, but I am not sure (for the usual reason, explained already in Note 2a); cf. p. 319 of [K 9] for references to other instances of that *petitio.*

(ii) In those notes to (13) Gödel is particularly critical of Turing's assumption that we do not have enough room in our heads as it were for our mental processes to be governed by an unbounded operator. This neglects growth of our — tacitly, finite — intellectual equipment. In other words, Turing's and Gödel's remarks may be regarded as jeux d'esprit corresponding to the sober distinction between — the behaviour of — bounded and unbounded operators; cf. a(i) above.

(iii) When Gödel mentioned (ii) to me in the 60's, I brought up the more radical objection, mentioned at the end of 3b (but also on p. 319 in [K 9]): the operators themselves may be non-recursive. His — I believe, considered, not offhand — view was that we know so little about the details that only very simple assumptions can be convincing. But here, in contrast to his reaction to other, apparently comparable cases reported at the end of Note 3, he rejected the thought that we may know too little for *anything* convincing. — He had no gratitude for small mercies like a(i) and (ii) above.

5. *Last sentences of Gödel's publications*. Remarkably many are defective. Here they are used mainly as pegs on which to hang sundry observations. But some serve also as memorable object lessons; for example, in (b) about reading too much into the printed word, or in (d) about the way a whole story can be lost when things get into print. It is relatively rare that impressions of such things can be checked against fuller details, and some readers may wish to do so. The digressions at the end of this long note are directed to such readers too.

(a) The best known example concerns (8) with a decision procedure for — the validity of — $\forall^n\exists^2\forall^m$ formulae of predicate calculus. Goldfarb [Go 1] has shown that, contrary to the last sentence of (8), the addition of = makes this prefix class undecidable (and thus somewhat exceptional). When doubts were raised in the 60's, I had no view about the truth of the matter, but suggested the following recipe for making a mistake: note that the full equality axioms follow from the atomic cases, and thus from universal axioms, and absorb them in one of the blocks $\forall^n$ or $\forall^m$ above (in other words, forget that a universal premise becomes existential in prenex normal form). I once even began to speak to Gödel about this, but got sidetracked by (my own) speculations on the circumstances that favour such mistakes.

(b) The finiteness, nowadays called compactness, theorem, at the end of (1) is formally correct, but defective in being unnecessarily restricted to countable sets of formulae. It came up on 2 occasions.

(i) I was struck by the fact that soon afterwards, in the note (3) on propositional logic, Gödel stated its result explicitly for arbitrary sets. Also I knew that he often took the opportunity of improving earlier formulations in later notes even if they were only tenuously related; cf. 2b on putting his undecidable sentence into the form $\Box p \to p$ or 3b on the absoluteness of formal computability in a note on speed up. As he himself described the matter he had first stated the propositional result for countable sets of formulae too, but found, on rereading the fortunately short note attentively, that the proof nowhere used countability. So he reworded the theorem, but was not interested enough in the generalization to look for parallels. Here it is to be remarked that, at the time, the general formulation might well have clouded the issue with worries about writing down uncountable sets of formulae (even though, realistically speaking, arbitrary countable, not necessarily r.e. sets are not very different in this respect). So the generalization might have limited the market for (1). Incidentally, the finiteness theorem was an afterthought that did not appear in Gödel's dissertation. — By contrast Malcev, who used the finiteness theorem in algebra, was not liable to trouble *his* public. The finiteness theorem for arbitrary sets was needed to get his algebraic conclusions in the (unrestricted) form usual in algebra.

(ii) Before I met Gödel I had simply assumed — cf. pp. 169-170 of [K 3] — that the finiteness theorem was consciously intended to answer the question: Given that the formal rules of Frege are complete for validity, what about consequence? the latter being defined for all sets of formulae (that one cares to consider). It certainly was not quite clear what a formal correlative might be. The finiteness theorem avoids any agonizing on this score. Gödel agreed of course, but did not remember thinking in these terms at the time.

(c) At the end of the first part of (9), $V = L$ (there called 'A') is described as 'a completion' of current axioms of set theory. Evidently this was out of tune with Gödel's song (and dance), starting a few years later, about the search for axioms valid for the full cumulative hierarchy or at least for far-out segments. If one has a particular notion in mind one speaks neither of 'a' nor of 'the' completion. What he did have in mind was coded in the terminology 'L' for: lawlike. At the time he toyed with the idea that L contained all legitimate definitions of sets. And he clearly changed his mind before he gave his lecture at the bicentennial celebration at Princeton (though, by Note 1b (iv), he was always reluctant to recognize, let alone, to enjoy any change of mind). Incidentally, the worry on p. 557 of (9) about the 'vagueness' of the notion of arbitrary set is removed in (12) by a reminder about the clarity of the notion *subset of*. — The next item contains the digressions announced at the beginning of this note.

(d) cf. also 4d (iii). In one of the later versions of the English translation of (13), Gödel attributes the last sentence to me. The details are more entertaining than the fact.

(i) By 4a I ignored his warning that he had treated his terms of higher type purely formally because I had already decided on an interpretation, for which the fan theorem was obviously valid; and I told him so a few days later. He remembered the remark when he came to write (13), but forgot to check whether it applied to his interpretation too. We spoke briefly about this. I pointed out in print, probably on several occasions, that the last sentence was not evident (without, however, mentioning the background), and there the matter rested.

(ii) During his final illness he brought it up, apparently worrying how to draw attention to my 'contribution'. The way he saw — or, at least, put — it was this. He regarded Spector's posthumous [S] as an important contribution (as he had already said in a PS); cf. 4e. In his view, given Spector's background — in particular, all he had learnt from Kleene's lectures about ordinary bar recursion and its relation to the fan theorem, and of course the idea of passing to all finite types in (13) — that last sentence was enough to trigger [S]. By (iv) below, Gödel's view about Spector's education was wrong.

(iii) But on this view, the next order of business was to find an appropriate wording for my 'contribution'. He first proposed to add 'for a slightly different interpretation.' The sequel was predictable. I asked if this was supposed to be a translation of 'eine gering abweichende Interpretation'; meaning of course his blunder discussed in 2a (i). The allusion had to be explained to him, and he was not amused. Obviously, it was not the person I had known for 15 years. I never asked him what wording he chose in the end.

(iv) Here is a sketch of some things I know about Spector's background before he embarked on [S]. When we first met at Cornell in 1957, where I had presented the material in 4c, he told me he was sick of Turing degrees, but also told me of his difficulties with the no-counter example-interpretation, which he had once presented in Kleene's seminar. At that time he concluded that it must be nonsense since it reduced arithmetic to Π^1_1 statements, without being contradicted; cf. 4c (ii). Well, Gödel's finite types, Π^∞_1, sounded worse still. I reminded him that it's not what you do, but the way that you do it, that counts; in particular, not the mere mention of the *language* of higher types, but the particular *properties* (or axioms) used, and we looked together at some striking examples. We kept in touch, and he visited me 1959 at Los Altos near Stanford. In the meantime it had occurred to me that I had derived the n.c.i. from Ackermann's version of Hilbert's ε-substitution method, involving in an essential way an order of priority. One had thus a relation to a principal element of Spector's logical background, Friedberg's priority method. In the meantime the popular article on Hilbert's program, reprinted in [BP], had appeared, with a reference to some *lurking lemma* in Ackermann's work. The phrase took Spector's fancy. In fact, he had studied the matter before the visit, and thought that he saw a lurking lemma (without putting it into words; in fact, till his death he said that he had employed that lemma as a principal trick in his functional interpretation of the negative translation of the axiom of choice; cf. p. 2 of [S], and 4e (i) for more on this). In short, his work grew out of a great deal of familiarity with ideas and results surrounding functional interpretations, helped perhaps by a few hints stemming from my own experience in this area, not least, the central place I had given to continuous functionals in my presentation at Cornell. — The next digression, on Gödel's, incidentally to me very congenial, literary taste will now be introduced (and concluded) by reference to Spector's paper. But it is a fluke that this kind of transition from (iv) is possible since the matter is general.

(v) Intended and discovered ambiguities. On p. 2 of [S] Spector quotes Gödel and Bernays as saying that bar recursion of higher type is just as evident as Brouwer's bar recursion (of lowest type). Spector himself felt encouraged, and, by footnote 2 of [S], so was I; my reservations in paragraph 2 of 4e developed later. I remember Gödel's glee when he pointed out that it could also be interpreted as follows: Brouwer's bar recursion is no more evident than Spector's generalization; cf. p. 159 of [RS] with a reference to de Gaulle. I did not bother to ask Gödel how he had intended the remark originally, in line with the view of intelligent literary critics about the 'life' of a literary product being — best regarded as — independent of its author's thoughts. Actually, Gödel's conversations were full of such ambiguities, and, once sensitized to this, one finds them also in his writings. For example, the parenthetical qualification in 'inhaltliche (intuitionistische) Überlegungen' of his lecture at Königsberg will mean to the intuitionistically indoctrinated reader 'and, hence, intuitionistic', but to the seasoned logician 'here, for once, intuitionistic'. Granted those intellectual reflexes attributed to young Gödel throughout this article, it is in the cards that this splendid ambiguity just came to him without any brooding. — Perhaps it is appropriate to add a couple of personal remarks. The first concerns ambiguities in my references to Gödel's intervention in choosing the title for Spector's paper. In 1.8 of footnote 1 on p. 1 the printer omitted the words 'by adding:' after the semi colon (not colon!), and I did not correct it. This puzzled at least one reader of p. 174 in [RS]. Here is the full story. In accordance with the views we had come to share about the consistency business, Spector's simple title was: *Provably recursive functionals of analysis.* Gödel did not find this exciting, and proposed the addition: a consistency proof of analysis. If I had known his Königsberg lecture at the time I should have quoted it back to him: cf. Note lb (i) above. But I didn't. Of course, I appreciated his flair for attracting attention, but my views about the sham of the consistency business have remained uncompromising. So, to water down his addition I proposed the further qualification 'by an extension of principles formulated in current intuitionistic mathematics', to which Gödel agreed, albeit reluctantly. The second remark is more speculative. Presumably, all of us who have a liking for — hearing or making up — ambiguities view them as a spontaneous game of hide-and-seek as it were; as in *cache-cache* and Talleyrand's or Fouché's *La parole a été donnée à l'homme pour cacher sa pensée.* (In contrast, for de Gaulle, cited above, it was not only a game.) But sometimes it seems to me there is a darker side to it, especially for those of us who have to do with foundational fundamentalists, notorious for their — cult of — literal-mindedness. The reaction to them is an almost inhuman coldness, viewing them as a different species, although one knows that many of them are worthy people. The game can be a relief, engendering an illusion of *complicité* with our own species somewhere out there.

6. *Some conversations on the proper order of priority* in logical research. In the 70's — when Gödel had become less 'formidable' in the words of Caroline Underwood, the secretary quoted on p. 160 of [RS] — he spoke and wrote rather freely of this matter, and not particularly convincingly; cf. Appendix 1. He spoke to me about it mainly after I had demonstrated my own interest spontaneously, cf. his view in Note 3a. Specifically, my statement in support of his election to the Foreign Membership of the Royal Society pointed out how Gödel's principal results were related to simple philosophical distinctions that others had ignored; cf. also p. 150, 1.21 of [RS]. Gödel's comments on that statement express the kind of pleasure mentioned in Note 2a, when others had by themselves come to views similar to his own. The following reminiscences give some idea of the way Gödel liked to muse about such matters in the mid sixties.

(a) He was of course much impressed by Cohen's results; not only because, as people often say, Cohen 'beat' him; after all, Gödel is on record as having been interested in the problem itself, in fact, as having considered the continuum problem fundamental; cf. 1.-13 to 1.-11 on p. 213 of [RS]. But, at least as he put it to me, there was more to it. He had thought that with a problem as fundamental as — he regarded — CH, the proper strategy was to reflect on the answer for the (or an) intended meaning, and then to translate it into formal terms. This had been his way for the constructible — or, by Note 5c, lawlike — sets; in more traditional terms, the transfinite extension of the ramified hierarchy, with simple types being replaced by cumulative types in the style of Zermelo. Already as a student Gödel had felt sure that Skolem's argument for defining elementary submodels of any infinite cardinality would establish the axiom of reducibility; cf. p. 198 of [RS]. The formal paraphernalia for converting this into a relative consistency proof was heavy only as long as he wanted to avoid the use of replacement, his first contact with axioms of infinity; cf. bottom of p. 196 of [RS]. Since Gödel had come to believe that CH was false for the full cumulative hierarchy the proper strategy would be to reflect on the latter, and to convert this reflection into a relative consistency proof for $\neg$ CH. No *general method* of constructing models would be needed. Cohen had provided such a method; cf. p. 201, 1. 15 of [RS]. NB. Gödel knew what a mathematical method was; *he* never used this word for the fixed point lemma at the end of 5e (i) nor for the enumeration of formal objects, that is, for the idea of Gödel numberings (at least, not when I used to see him) nor even for the formal definition of the arithmetic operations derived by that enumeration from those on the formal objects.

(b) He was quite aware that his own attempts in the forties to prove the independence of AC did not employ the strategy of (a), but a reinterpretation of the logical particles, in clumsy syntactic terms to boot. It is fair to say that the idea behind it is very well expressed by means of boolean-valued models. (i) As has often been noticed, in [Co 1] Cohen professed to be a formalist after he had used models in his independence proofs successfully, and had given a feeble relative consistency proof at the end of [Co]. Gödel's way is more congenial to me; he bandied 'platonism' about after troubles with the syntactic methods just alluded to, and with interpreting his primitive recursive terms of finite type purely formally; cf.4a (iv). — The next item is touching, but seems to have a moral too. (ii)

On one occasion he mused about not having published those syntactic reinterpretations; people might have misunderstood him to mean that those were the right interpretations. I never misinterpreted his remark to concern his conscious reason at the time, but rather abstract possibilities; in particular, situations where attention is drawn away from a potentially fruitful problem by an ad hoc or otherwise unsatisfactory, albeit correct solution. (A formal error often draws attention *to* the problem; cf. Note 5a.) Individuals of a certain temperament do, in fact, worry that a better solution may not be so widely acclaimed as a first solution of problem. (Other temperaments derive confidence from knowing the answer or like the idea of doing better than a well known author, especially if the latter has taken the trouble to treat the problem in print, and so forth.) With *current* mass activity Gödel's simple view of triggering chains of events may well apply to *somebody*, though not in the cases to which he applied it specifically. For example, by Note 5d (ii), not to Spector, nor to the (published) monograph (10) with an introduction preoccupied with legalistic precautions; cf. (b) on p. 197 of [RS]. As already alluded to at the beginning of A1c it is quite remarkable how little work was done on the constructible sets in the 40's and 50's, though, as Jensen has shown, there was a lot more to discover about them.

(c) *Reminder:* Gödel's early exercises on i.l., described in section 2, are at the opposite extreme to the order of priority, in (a) above, that he had come to advocate later. The style of (13), and especially of the notes that he added for the English translation, serves as a foil, cf. 5f. — *Personal remark.* Partly because of the musings in b (ii) I began to record my expectations of various projects, especially, in periods of consolidation; by p. 140 of [K7] with the explicit purpose of checking them against later experience.

7. *Further conversations on titles, terminology and other expository devices* (beyond those that have come up before, for example, at the end of Note 5). Our ordinary view of human nature, as described at the beginning of A 2d, requires a kind of causal interpretation of the anecdotes below; involving motives (rather than 'mere' reasons). I do not often find the view compelling; certainly not here. And I have nothing to contribute to such interpretations. As is clear from the preceding Notes, Gödel often spoke of expository tricks to be found in his publications and I never bothered to

ask if he had had them consciously in mind at the time or meant their general relevance (to repeat: not cause and effect, just as one does not ask for cause and effect when relating equilateral and equiangular triangles). Below, the anecdotes will be used to enrich the usual view of Gödel's work, and also of its relation to later work in logic. The Note proceeds by easy stages from the sublime to the ridiculous, with a digression at the end.

(a) Gödel had a strong dislike of the slogan 'the meaning of a theorem is its proof', which was current in his student days; it fits well the boring statements of theorems in the constructivist literature (and of course its interpretation of the logical particles as, literally, operating on proofs). At least in conversation with me he insisted that only results be mentioned since their pattern might be obscured by the proofs; not only mine but also by nice ones done by others. What about information contained in proofs, but not stated in the theorem? For Gödel the first order of business was to state elegant and memorable theorems. Afterwards people can look at the proofs for additional information of interest to them. His practice followed this principle, very much in contrast to Herbrand and Gentzen who — before and after the completeness and incompleteness theorems — used all purpose expressions (Verlegenheitsausdrücke) like *théorème fondamental*, resp. *Hauptsatz,* without any explicit indication of what made those theorems so 'basic'; but cf. A 2b (i). Some 10-20 years after them I attempted to find concepts more adequate for expressing at least some of the additional information provided by their kind of metamathematics; for example, functional interpretations rather than consistency, conservation rather than relative consistency, and many more besides. I do not think that Gödel felt comfortable with those concepts though by now they are familiar enough to be considered memorable. A *blind spot*: during the period of our frequent conversations I had not yet realized clearly enough — for a rewarding discussion — that contemporary mathematics has its own response to the slogan above: Find the concepts needed to state theorems that express the meaning of a proof. ('Meaning' in the sense of — what is called crudely — essentials.) Generalizations, tacitly, in terms of skillfully chosen concepts, often do just that; cf., for example, section 1 in the review of [F]. This is philosophical progress for the kind of philosophy meant in section 1 above; not least, for the face-saving powers of the low-keyed mathematical style. Nobody feels ashamed about having to search for some — or even a particular kind of — generalization. But many feel ill at ease when they do not know what is essential about a proof or what it means.

(b) The following specific conversations on titles and terminology touching topics of the present article seem pretty typical. (i) Soon after we first met, Gödel made fun of the title of [K]. The content appealed to him; a twist on his incompleteness theorem that would have been perfectly accessible in 1930, with an undecided sentence U in Δ_2^0, but described in [K] in terms close to what would today be called: of degree $\leqslant 0'$. The proof leaves open the parameters (systems and/or codings considered) determining whether U is true or false; cf. [MS] for partial results. As noted later the twist yields what is still the most 'logical' model-theoretic proof of Gödel's *second* theorem; cf. pp. 862-863 of Smorynski's article in [B]. Gödel found it odd that one could be clever enough to find the results, but not a sensible title. Obviously, one does not spoil such a remark in conversation by boring analyses of the circumstances. But here it seems worth adding that — I thought — I had not achieved one of my aims in [K]; specifically, of a consistent formula without any recursive model (in fact, even without any recursive valuation for its *atomic* properties). What I had actually proved was that GB including the axiom of infinity — or, more pedantically, its Skolem normal form — had no such model. And at the time I did not know that GB had any model at all! I did not know the cumulative hierarchy; cf. [K 12] for an account of the comedy involved. Nor did I have the experience needed to find the words appropriate in such circumstances. — (ii) Gödel complained, equally pertinently, about 'basis' in 'basis theorem'. The notion is popular enough, at least, after Kleene's exposition [Kl], but probably not the way I have always looked at it. It was to extract the sensible side of Ockham' s razor; without the absurdity of supposing that things do not exist when they are not needed (to handle the phenomena and problems that have so far struck us). — (iii) A footnote on p. 225 of the Heyting Festschrift in Compositio, Vol. 20 describes Gödel's objections to 'absolutely free' for — what, following his proposal, are called — lawless (choice sequences). Several points are to be added. First, at the time, I did not connect the proposed terminology with his code meaning for 'L' in Note 5c; the pair 'lawlike' and 'lawless' is not only catchy, but easily translated: gesetzmäßig and gesetzlos. (I take it he saw no need to use a code like 'L' here because I was not going to be reticent about the meaning I intended anyway.) Incidentally, today I prefer the terminology 'open data' because it expresses explicitly the relevant enrichment. Secondly, Gödel was touchingly pleased by the innocent successes of socalled informal rigour applied to lawless sequences; in other words, of the strategy he advocated in connection with sets; cf. Note 6a. For example, the axioms for open data or the decidability of extensional equality for lawless sequences are recognized by inspection. Certainly so far, the notion of lawless sequence has been quite comparable to the primitive notion of set as a *source of axioms* (though not as a scientific tool); and much more rewarding in this respect than the primitive notion of effective rule; cf. 4d (ii) on the contrast between the latter and that of set. Thirdly, at least as I see things now, the principal philosophical interest of lawless sequences derives from the object lesson they provide for the topic of natural languages, elaborated near the end of section 2. To repeat, a precise and elegant development is perfectly possible, but paraphrases are just as effective. Finally, at least so far, nothing much has come of Gödel's great expectations for lawless sequences of higher type objects; cf. the second paragraph of 4e.

(c) The items (i) and (ii) below are extreme examples of the kind of musing — on 'manipulating' the reader — that Gödel enjoyed very much. But for a proper perspective it seems necessary to get a couple of generalities out of the way. Thinking about such manipulation — from advertizing to formal education — involves not only highly publicized 'normative' elements about desired effects, but costly empirical elements in assessing actual effects; 'costly' because either many people are involved or, as in the case of education, only long term observations are of much use. So, on the negative side, it's hard to establish any conclusion (and therefore often difficult to refute silly opinions). On the positive side, as already mentioned in Note 6b (i) about current mass activity, Gödel's short term view on getting immediate attention by at least a few able people is perhaps more effective than it used to be; the activity will lead to a kind of natural selection, not only of concepts; cf. the story of determinacy on p. 208 of [RS]. With relatively few exceptions silly, ideologically inspired shibboleths have been quietly dropped in the last 15 years. For example, one no longer speaks of 'deriving consequences from the axiom $V = L$' when in fact one is proving properties of L (which generally have those consequences as trivial corollaries, but not conversely). Slowly one is beginning to talk of the rate of growth of bounding functions for Π_2^0 theorems, and not only of formal independence (when in fact one has proved independence from all true Π^0_1 theorems; cf. A1b (iii)). The progress achieved in this way becomes spectacular if contrasted with one of the exceptions; especially with [NH], which revives remarkably many thoughtless first impressions that were corrected by logical research in this century. (The drivel about the fundamental character of relative consistency is put in its place in A1 c.) — Here are the two promised titbits, both about the incompleteness paper.

(i) When I once mentioned to Gödel that the introduction to his incompleteness paper was fully convincing, he agreed, but thought that the masses of formulae in later parts had served to avoid futile discussions about any relations between his work and Finsler's [Fin]. Nobody looking at both would even dream of worrying about such relations. Several things should be added. First, a concise formulation of those relations needs, as so often in such cases, considerable familiarity with the subject; cf. Note 8b (iii), but also the footnote on p. 169 of [RS] about a quite closely related business with Zermelo.[7] So it is not sensible to have outsiders worry their

[7] *Addition* to the top of p. 154 of [RS]. The article of Olga Taussky in this volume contains her view of a meeting between Gödel and Zermelo at Bad Elster in 1931. Whatever — illusion of an — understanding may have been reached at the time, it is certainly no longer expressed in Zermelo's letter of 7th October 1931 to Reinhard Baer (p. 43 of this volume). The perhaps more interesting question, whether Gödel's brilliant exposition in his letter of 12th October 1931, referred to in [RS], was of more help to Zermelo than the apparently cozy walk near Bad Elster, is of course not settled by the letter of 7th October (written while Zermelo was waiting for an overdue reply to his earlier letter to Gödel). But to repeat a theme of this article that cannot be repeated too often: I know too little of Zermelo's personality, apart from the fact that it is very alien to me, to make further speculations on the details above rewarding. However, there is another side to the matter which can be viewed as an instance of an almost universal temptation.

As shown convincingly in [Z], and elaborated in [Z1], Zermelo had been, or at least, had become convinced, quite independently of Gödel's incompleteness theorem, that formal systems were inadequate for understanding logical reasoning. (Reminder: formal systems have recursive, that is, hereditarily finite rules, and [Z] concerns the logic of the Infinite.) Of course, Zermelo's view was shared by the silent majority. But in contrast to the latter he said something about his view (in [Z1]).

In terms of Walpole's distinction, in Zermelo's place a man of thought would have used Gödel's theorem to strenghten the case against formal systems. They are

heads over relations to [Fin] (even if academic etiquette requires some kind of reference; tacitly for insiders). Secondly, on the negative side as it were, the price paid for introducing all those formulae was high; it is, at least, to some extent, the superstition that Gödel's proof is subtle. Last, but not least, as I realize now, the general presentations of the incompleteness theorems in Hilbert - Bernays (with Gödel's help, cf. 3b) are not much longer than Gödel's original introduction; of course, the verification of the general conditions by a specific system may take some effort. This is a, philosophically, more satisfactory balance; for another abstract formulation, cf. the last paragraph on p. 171 of [RS] (where on 1-10 '(**) in F' should be replaced by '(**) *or* F'.

(ii) As is well known, the title of the incompleteness paper suggests a sequel, but none has appeared. I never asked Gödel about the general circumstances, which might allow one to judge to what extent it would be sensible to speak of 'causes'. Anyhow he volunteered a view on the matter. If there had been massive and systematic misunderstanding of the paper, Part II could have been used to give a full statement and proof of the second theorem, so to speak as its principal purpose; and some of the — actual, not merely imagined! — misunderstandings would have been corrected incidentally. Viewed this away, my leaving Gödel's work on i.l. out of [RS] has turned out to serve a similar purpose. I have used the present article — not so much for correcting misunderstandings, but — for reiterating certain points of [RS]. This purpose certainly did not occur to me when I started on [RS], nor even on this article, which sets out in the opening paragraph my (conscious) reasons for neglecting i.l. in [RS]. — We now come to the digression announced earlier in this Note.

not merely inadequate in the obvious sense already mentioned, of 'embracing all valid methods of proof', but even w.r.t. provability (of true Π_1^0 sentences). In contrast Zermelo, apparently a man of feeling, reacted to Gödel's theorem as if any contact with formal systems — even by way of negative metatheorems about them — was bound to be pernicious, like witches and other creatures of the devil.

Gödel himself is of course the perfect counter-example to such fears; cf. p. 210 of [RS]. But it should not be assumed that the man of feeling is statistically all that wrong about his many fellowmen (of feeling).

(d) Some ideas of Gödel about logicians with or without his flair for flashy formulations. (i) He often called Gentzen a better logician than himself. Obviously, Gentzen was not more interesting; but his results were not in the air (nor on the surface: it took a long time to see convincing implications). (ii) Gödel had such a high regard for Kleene's contributions that his wife complained about the Institute for not making Kleene a professor there. In particular, always appreciative of a twist to his incompleteness theorem Gödel talked with relish of the formulation in terms of disjoint r.e. sets that are not recursively separable. (iii) By 1957 he had so much confidence in Dana Scott that he said he expected Scott to prove soon the formal independence of CH (from AC). And in fact, Scott contributed fairly soon to Gödel's favourite proof (by use of boolean-valued models). (iv) When he told me how much he liked Takeuti's contributions to proof theory I asked for a summary. Knowing my logical tastes Gödel said instead that Takeuti had so much talent for seeing through complicated combinatorial situations that he did not need the kind of abstract view I wanted (and that he, Gödel, could not formulate adequately such a view). With due regard for conversational licence I'd say much the same after my experience in preparing the joint paper [KT 1]. — Perhaps (i)-(iv) balance a little the embarrassing exaggerations in Gödel's interview with Time Magazine after Friedberg's solution of Post's problem. (As — bad — luck would have it, Gödel went out of his way to say that this kind of achievement happened once in a generation, more or less when Mučnik's paper was in press.)

8. *Afterthoughts*, the bulk of which could (and should) have been integrated in earlier notes.

(a) Complements to Gödel's musings about (logical) research, especially in Notes 6a and 6b about CH.

(i) The principle in Note 6a jars a bit with Note 3a (iv) about trying harder. But if this does not mean harder mathematics, what should it be? — Of course, Gödel may have read the motto of the car rental firm Avis (we try harder) on the day of the conversation reported in Note 3a (iv). Then one would not attach much weight to his precise choice of words.

(ii) A certain see-saw in the sense of 2a in fine is to be expected since by Note 6b the 'principle' in Note 6a reflects Gödel's own failure with the opposite strategy. But the see-saw continued into the fifties with his encouragement of work on intermediate r.e. Turing degrees for new ideas on CH, reported on p. 159 of [RS]. This suggests at least a couple of thoughts:

(iii) On the positive side there are certainly objective relations between (perfect set) forcing and, for example, some (of Spector's) methods in the theory of Turing degrees.

(iv) On the negative side — and perhaps more interestingly — Turing degrees of r.e. sets require at least a certain measure of symmetry between (the complexities of) the sets and the mappings considered; in contrast to CH for the full cumulative hierarchy; cf. top of p. 213 of [RS]; 'certain measure' because — as some of us sensitive souls have complained — r.e. sets and their complements are severely asymmetric (w.r.t. proofs of membership), while Turing reducibility is not.

(b) Tricks of the memory, expecially during Gödel's final illness.

The subject has fascinated many; most recently, the writer Marguerite Duras (*L'amant*) who made a point of recording her memories as they presented themselves to her, though it would have been easy to correct many quite obvious errors, for example, chronological discrepancies. From a solemn point of view like Freud's, one would look for specific causes of each slip; a kind of hybris to suppose that they can be found even approximately without knowledge of memory structures; cf. A2d (ii) on the inner life of planets. (As always, there are exceptionally favourable circumstances; cf., for example, d (i) and (ii) below.) — Here we treat those tricks quite differently, along the lines of Note 7c (i) about suppressing references to earlier literature.

A particular trick of memory involves Gödel's recollection — admittedly, in his less 'formidable' period (cf. p. 160 of [RS]) — that his work on CH had nothing to do with Hilbert's sketch in the paper *Über das Unendliche,* already referred to in footnote 4 in 4d. The facts are plain enough.

(i) In notes for a lecture at Brown University (1938) in the Nachlaß Gödel explicitly says that he had recently discovered a presentation of the work that was closely related to Hilbert's sketch.

(ii) At the end of his review in the JSL of (9), the genuinely cautions, not merely calculating Bernays says the same thing; and, after all, anybody competent can verify some objective relations.

(iii) Granted the similarities between Hilbert's sketch and Gödel's proof, the differences are surely much more impressive and consequential. As in his ε-substitution method, which was a main topic of Hilbert's paper anyway, he thought of the ordinal-theoretic functions involved as 'essentially' finitistic. For Gödel's (successful) work it is essential to think of them — in modern terminology — as α-recursive (for constructible cardinals α).
Remark. By p. 202 of [RS], in 1936, and later in the 70's, Gödel himself felt that he had provided a first step towards the kind of further 'collapse' to finite structures needed for Hilbert's program. But before Girard's, admittedly, blatantly transfinite iterations of limit processes applied to dilators, there was hardly a hint for achieving that kind of collapse. (The collapse of cardinals into recursive ordinals in Bachmann's hierarchies was not enough, at least not for me, to inspire confidence.)

Thus for Gödel's primary concern, already noted at the beginning of Appendix 1 — of making his work accessible to a wide audience, and so necessarily without much background — it is certainly most appropriate to draw attention *away* from the similarities to Hilbert's sketch since a *correct* appreciation of them demands substantial background knowledge.

All this looks bad from a more solemn point of view. First, as in Note 7c (i), we have a conflict with academic traditions about acknowledging priorities. But since human beings, and the situations in which we find ourselves, differ, the best we can hope for from traditions, including laws, is that they are appropriate in many cases. Besides, academic traditions are not primarily concerned with a wide market.

Secondly, Gödel's later views of the facts occur, if I remember correctly, in letters (meant for posterity to boot), not in a casual conversation.

Instead of — or after — enjoying the thrill of indignation, let us take a look at the solemn view.

(c) Documentation, for example, by 'official' publications, but also letters (including those above intended for posterity) *versus* conversational anecdotes. We begin with a general reminder.

(i) In political history, letters and secret memoranda or tape recordings have proved to be useful, often more so than relying on public pronouncements; at least after extensive sifting, with due attention to the temperaments of the authors. Certainly, *one* element that contributes to reliability is that the documents are produced by many different people, and affect even more. Recriminations after the event automatically generate new material. *This* element is not present to the same degree in scholarly doings, and further restricted by academic etiquette.

A principal assumption of the history of ideas is that, in contrast to politicians, scholars do not try to manipulate their public. But even when this is true — and Note 7 serves as a *caveat* —, the scholars, and above all the pioneers, often simply do not know how to say what they know; cf. Note 7a. So, realistically speaking, as a reflection of their thoughts,words of scholars, whether written or spoken, may be less reliable than those of politicians.

Without exaggeration, when the history of ideas apes the kind of documentation familiar from ordinary history it is liable to become a parody. All this is of course in keeping with the suspicion of the history of ideas among the silent majority of historians. — Here it should be added that many professional historians are also unsympathetic to the use of history for political rhetoric (which,for the record, I like; just as much as for entertainment despite Nietzsche's diatribes in *Unzeitgemäße...*). In mathematics such rhetoric tries to convey a view, for example, on the relative promise of different methods, by presenting a suitable selection of historical titbits, including of course hopelessly false starts, according to that view; cf. also Comment (ii) on p. 233 of [KM]. The historians' antipathy goes well with the fact that pseudo-history can be quite as effective rhetorically as real history, provided it catches our imagination.

The following points about letters, resp. publications are quite down to earth.

(ii) Bertrand Russell complains in the introduction to — the *second* edition of — his book on Leibniz that the latter never wrote a *magnum opus* of universal interest; in Russell's view because Leibniz wrote too many letters to princesses; presumably — if there is to be a conflict at all — relying too much on references to specific interests of those ladies.

(iii) Depending on the temperament of authors formal publications, expecially on intimate subjects like broad views, are not always as different from letters as (ii) suggests. Speaking from personal experience while writing this very article, I caught myself addressing particular readers; sometimes dead ones, sometimes present company. Even if I am a bit exceptional in this respect, my aberrations probably merely magnify a widespread phenomenon. The same surely applies to the following anecdote about an, admittedly, exceptionally simple-minded person.

(iv) By a fluke I had recently the rare luck of learning the interpretations by that person of some letters I had written to a friend. Phrases that just happened to have caught the ear of the friend in conversation, were given portentious (and always inappropriate) meaning. Others that were of genuine concern to the person were completely overlooked. — Of course, this does not mean that it is totally unrewarding to read letters not addressed to one; but it underlines the unusual skills needed to guess the intentions. Again this does not mean that an interest in intentions is illegitimate or, perhaps, logically 'senseless'. It just isn't rewarding if the chance of anything beyond a very rough general idea is slight.

For balance, here is a 'positive' sample.

(d) Settling an historical question by looking at documents: how Gentzen reacted when Gödel and von Neumann criticized the original — posthumously published — version of his consistency proof for arithmetic [Ge]. The general background is well known; cf. for example pp. 202-204 of [K 13], where, on p. 203, the following harmless sounding question was asked:

How did Gentzen get from the criticized proof to the first published version?

(By the letters referred to below, he knew that the criticisms were totally mistaken.)

In fact, two — objectively perfectly sound — possible paths were suggested. It turns out that the *question was mistaken*. Gentzen dropped the criticized version, and simply went back to an earlier idea. All this is available in letters and postcards left by Bernays to the ETH, written by Gentzen in autumn 1935 to Princeton (where Bernays was in close contact with Gödel after their transatlantic crossing mentioned already). What is remarkable here is the lucky combination of circumstances that led to the answer.

(i) Bernays himself had spoken to me about the criticism in the sixties. In response I observed that even if the fan theorem were used, as was alleged, the consistency of arithmetic could not be proved since the addition of the fan theorem is conservative; cf. the review of [KV]. The conversations took place in the very room where Bernays kept his correspondence, but he had forgotten the relevant letters when I asked him about such things.

(ii) When I once asked E. Engeler at the ETH about correspondence between Gödel and Bernays while I was preparing [RS], I also saw a folder with Gentzen's correspondence. So on the off-chance that I might find something relevant for Gödel's obituary I asked Engeler later to send me copies; cf. also the long review in Zbl. 501 (1983) 21-24, #03045 concerning the Russian translation of [K 13] among several papers.

Clearly conversation with one of the participants — Bernays, not Gödel who never mentioned the episode to me — was needed to draw attention to the general area. But written documents were useful to supplement personal memories.

At this point it might be tempting to speculate on Gentzen's character; perhaps on his indifference about communicating an obviously fruitful idea — of using functionals (of lowest type) — in metamathematics, in other words, of 'defending' his proof; or on his self confidence in not being overimpressed by older and well-recognized people like Gödel and von Neumann. But I at least do not know enough about Gentzen to have any confidence in my ability to arrive at worthwhile conclusions about his character, even with the benefit of many stories that Witt has told me of his friend Gentzen.

The snippets above bring to mind a further afterthought about Gödel's later conduct, (objectively) related to them, and — at least, to me — memorable and congenial.

(iii) *Wenn schon, denn schon*; cf. 3c(ii). At first Gödel, like von Neumann, was ill at ease with Gentzen's use of functionals, albeit of *lowest type*. But when Gödel returned to the subject, about 5 years later, he used *all finite* types; cf. section 4.

(iv) On the negative side as it were, Gödel's attempts to assimilate bar recursion of higher type to that of lowest type, reported in Note 5d (v), have some of the flavour of the mistaken criticism alleging the use of the fan theorem in Gentzen's argument. The general point, already stressed in the digression in 4e (i), was that merely an *abstract combinatorial principle* was at issue, and so the complexity of the definition of the trees or, more generally, of the ramification schemata should not affect the evidence of the principle. For example, induction is valid for a predicate provided the latter is well defined at all (even though the formal or proof-theoretic strength must be expected to depend on the formal complexity of the definition; after all, this is one element of formal incompleteness, and the weakness of proof-theoretic strength as a measure of mathematical difficulty).

Here is one of the rare cases where the rhetoric about proof and truth is not empty, at least, if — something like — the idea of fully analysed proof is kept in mind at all. To recall 4e (i):

If the elements of a tree are defined by a complicated condition C (on finite sequences) with terminal nodes T then fully analysed proofs of

well foundedness, that is, $\forall\alpha\ [\forall n C(\bar{\alpha}n) \rightarrow \exists m T(\bar{\alpha}m)]$

just need not look remotely like such proofs of $\forall\alpha\exists m T(\bar{\alpha}m)$ There is nothing subtle about it; for given α and n the proof may use recondite properties of proofs of $C\,(\bar{\alpha}n)$. If the values αn are chosen from a decidable species there is no need-in-principle to use recondite proofs of assertions $C\,(\bar{\alpha}n)$ or $\neg C\,(\bar{\alpha}n)$

Gödel himself paid little attention to this aspect. As explained already, his primary concern was to bypass the matter by use of the primitive notion of effective rule of finite type, for which $C\,(\bar{\alpha}^o\ n) \vee \neg C\,(\bar{\alpha}^o\ n)$ should hold too. This was in accordance with one of his favourite maxims: if you work on a subject you might as well think that there is nothing else in the world. Readers with experience of monogamous relations probably see the full inwardness of the maxim.

(e) A word on reading the present article. For the view described as 'primitive' at the beginning of A 2d the style of this article is sufficiently unconventional to raise problems; for example, about my motives (in biblical terms: *les pères nus*, and the like; cf. Genesis 9,22). Naturally, I must leave those problems to others who see something of interest in them. I regard them as profoundly mistaken, no less than my question about Gentzen in (d) above. But for what they may be worth, here are random observations suggested (to me) by some of these unconventional features.

(i) Objectively, I am most impressed by the efficiency of a conventional style of exposition. Just being conventionally methodical, repeating standard definitions, and so forth, some totally uninspired authors can communicate material that more gifted people use to excellent effect. Applied to exposition this realizes the (democratic) ideal of progress formulated by Leibniz in connection with the calculus: '...ermögliche der Mittelmäßigkeit Probleme anzugreifen, die bisher nur den Hochbegabten zugänglich gewesen'; naturally, leaving it to the more gifted to choose the problems and to know what to do with the solutions. Here it is to be added that, as Littlewood points out in *A mathematicians's miscellany*, the current average level of exposition is an achievement of this century; in his youth it was taken for granted that just reading an article required previous research experience. — But, as far as I am concerned this appreciation of a methodical style is abstract. Already Ovid touched such problems of discipline: video meliora proboque; deteriora sequor. Perhaps, (ii) below 'explains' my lack of discipline, even if it does not 'excuse' it.

(ii) When writing this article I had [RS] very much in mind, and some reactions by the — absolutely, very small — audience to which it was addressed. One complaint, incidentally by the reader to whom it gives me the greatest satisfaction to communicate a thought I like, was that it is not easy to leaf through [RS]; hence the many divisions, subdivisions, notes etc. in the present article that can be read independently. On the positive side, at least for *all* the relatively many readers who commented on the matter to me, the most unconventional twist of [RS] had precisely the desired effect. (I mean of course the introductory fanfare about Gödel not having invented mathematical logic.) All those congenial to me — told me that they — looked immediately at the final paragraph for the so to speak compensating twist; all those of a different species in the sense of Note 5 *in fine* — told me they — were ill at ease about the fanfare or even misinterpreted it by overlooking the fact that *all* of Gödel's famous work in the thirties had been prepared for him; the problems he solved had been formulated and *made* famous, mainly by Hilbert! This encouraged me not to suppress my stylistic reflexes in the present piece (and confirmed my leaning towards de Maistre and against Condorcet in their dispute about such matters).

(iii) As stressed repeatedly throughout this article, problematic points in Gödel's work are given prominence. This is unusual, at least, on occasions like the present symposium. But I see it as a corollary of a specific and of a quite general fact. By and large, Gödel's expositions have been so effective that his unproblematic contributions have found their way into texts (cf. top of p. 192 of [Sa]), and — this is the general point — usually in improved form. At least statistically, the tradition of going back to the sources, so often appropriate in literature and the arts, cannot be expected to be equally effective in the sciences including mathematical logic, where 'progress' has a pretty clear meaning; cf. the early part of section 2, and the end of Note 8b (iii) on measuring the success of traditions. In view of differences between subject matter, not to speak of authors, only the short sighted would expect to rely equally on 'digging' painstakingly into sources for all aspects of all intellectual activities; even in connection with their history (cf. the end of A2), let alone, their exposition.

After these generalities we conclude with the following quite specific, even personal notes.

(iv) What is conventionally regarded as a sin — and, quite objectively, can indeed cause trouble for many people concerned; cf. Note 3b (ii) — often does not trouble me at all. For example, as I see the long list of defects, not only in Note 5, I am particularly impressed by their quality. Even today they continue to suggest worth while reflections, and much more so than many a tame perfectly sound contemporary publication. (Of course, Gödel's defects are not recommended to the rest of us: they separate the men from the boys.) Besides, at least to me, those defects related to his carelessness present a most welcome relief from that would-be philosophical constipated 'precision' (out of all proportion to background knowledge) that had always repelled me in some of Gödel's popular writings; cf. p. 158 of [RS][8]. Specifically, it is a relief to think of that style as a pose to impress philosophers (cf. [RS], p. 204, l. 18-19), even if in fact it only attracted philosophical cripples; rather than as coming from the heart or, in one of his favourite phrases, employed mit Lust und Liebe.

(v) To judge by past experience, I am obviously disturbed by seeing a picture of somebody I knew well that conflicts with my own memories; for example, when I first saw the selection by Wittgenstein's literary heirs among his remarks on the foundations of mathematics I described it as 'a surprisingly insignificant product of a sparkling mind'; more than 20 years later I had the good luck to find a documentary correction; cf. [K15].

In Gödel's case such things as (13), which, by 4c and 4d I continue to regard as a gem, are, of course, central to my picture of him; but such lists as in Note 5 belong to it, too. Together they do a little, at least, for me, towards balancing the singularly mediocre Gödeliana mentioned early on in this article.

(vi) A final word on 'straight', in other words, formal errors such as the one in Note 5a. Sure, a more cautious person than Gödel would *either* not have been tempted to put into print an ill considered answer to an — as it has turned out — unexpectedly interesting question *or* would have employed one of the standard academic conventions for avoiding offhand (formal) mistakes; by making equally ill-considered conjectures or most simply, by asking a question.[9] But objectively speaking, not without loss!

[8] Afterthought. I can't stand the brash article (12) on Cantor's CH either; cf. again p.158 of [RS], but also p. 212 (b), and above all p. 197 (c). The oversight mentioned there, about judging CH by its — non existent — arithmetic fruits, is not exactly in a last sentence. It was certainly not offhand, but the result of an ingrained blind spot, as explained loc. cit.

[9] In this connection it should be remembered that thoughtful mathematicians are sensitive to the abuse of the word conjecture (for off-hand questions), for example, A. Weil; cf. vol. 3, p.454 in his *Collected Works.* The degree of thoughtlessness involved in this abuse, in particular, w.r.t. evidence for such 'conjectures', is

Just think of the joy that our fellow brethren, who perhaps do not have many other joys in life, derive from having discovered an 'actual' error by Gödel. Again, making such errors cannot be recommended to the rest of us: their discovery would give less joy.

as staggering as anything (I know) in the philosophical literature, albeit in less pretentious language. Thus 'evidence' is bandied about without a moment's hesitation over the warnings in elementary texts on statistics about minimal precautions needed before such talk is profitable at all.

REFERENCES

(B) Barwise, K.J. (ed): *Handbook of mathematical logic,* North-Holland, 1977.

(BP) Benacerraf, P. and Putnam, H. (eds.): *Philosophy of Mathematics. Selected readings,* Prentice Hall, 1964.

(Be) Bezem, M.: "Strongly majorizable functionals of finite type. A model for bar recursion containing discontinuous functionals", Journal of Symbolic Logic 50 (1985), 652-660.

(Bu) Burgess, J.P.: "The completeness of intuitionistic propositional calculus for its intended interpretation", Notre Dame Journal of Formal Logic 22 (1981) 17-28.

(C) Church, A.: "An unsolvable problem in elementary number theory", Amer. Journal of Mathematics 58 (1936), 345-363.

(Co) Cohen, P.J.: "The independence of the continuum hypothesis", Proceedings of the National Academy of Science 50 (1963), 1143-1148 and 51 (1964), 105-110.

(Co 1) — "Comments on the foundations of set theory", Proceedings Symp. Pure Math. 13 (1971), 9-15.

(Cr) Craig, W.: "Boolean notions extended to higher dimensions", pp. 55-69 in *The Theory of Models,* ed. by J. Addison, Leon Henkin, Alfred Tarski, North-Holland, 1965.

(DG) Dreben, B. and Goldfarb, W.D.: *The decision problem. Solvable cases of quantificational formulas,* Addison-Wesley, 1979.

(Fin) Finsler, P.: "Formale Beweise und die Entscheidbarkeit", Mathematische Zeitschrift 25 (1926), 676-682.

(F) Fitting, M.C.: *Fundamentals of generalized recursion theory,* North-Holland, 1981; rev. Bulletin of the Amer. Math. Society 13 (1985), 182-197.

(Ga) Gabbay, D.: "A note on Kreisel's notion of validity in Post systems", Studia logica 35 (1976), 285-295.

(Ga 1) — *Semantical investigations in Heyting's intuitionistic logic,* Reidel 1981.

(GH) GANDY, R.O. AND HYLAND, M.: "Computable and recursively countable functions of higher type", pp. 407-438 in Logic Colloquium '76, ed. by R.O. Gandy, J.M.E. Hyland, North-Holland, 1977.

(Ge) GENTZEN, G.: "Der erste Widerspruchsfreiheitsbeweis für die klassische Zahlentheorie", Archiv für mathematische Logik und Grundlagenforschung 16 (1964), 97-118.

(G) GOAD, C.A.: "Monadic infinitary propositional logic: a special operator", Reports on Mathematical Logic 10 (1978), 43-50.

KURT GÖDEL

(1) GÖDEL, K.: "Die Vollständigkeit der Axiome des logischen Funktionenkalküls", Monatshefte für Mathematik und Physik 37 (1930), 349-360.

(2) — "Einige metamathematische Resultate über Entscheidungsdefinitheit und Widerspruchsfreiheit", Anzeiger der Akademie der Wissenschaften in Wien 67 (1930), 214-215.

(3) — "Eine Eigenschaft der Realisierungen des Aussagenkalküls", Ergebnisse eines mathematischen Kolloquiums 3 (1930-31), 20-21.

(4) — "Eine Interpretation des intuitionistichen Aussagenkalküls", Ergebn. math. Kolloq. 4 (1931-32), 39-40.

(5) — "Zur intuitionistischen Arithmetik und Zahlentheorie", Ergebn. math. Kolloq. 4 (1931-32), 34-38; rev. Fortschritte der Mathematik 59 (1933), 865-866.

(6) — "Zum intuitionistischen Aussagenkalkül", Ergebn. math. Kolloq. 4 (1931-32), 40.

(7) — "Über die Länge von Beweisen", Ergebn. math. Kolloq. 7 (1934-35), 23-24.

(8) — "Zum Entscheidungsproblem des logischen Funktionenkalküls", Monatshefte für Mathematik und Physik 40 (1933), 433-443.

(9) — "The consistency of the axiom of choice and of the generalized continuum-hypothesis", Proceedings of the National Academy of Sciences U.S.A. 24 (1938), 556-557.

— "Consistency-proof for the generalized continuum-hypothesis", Proceedings of the National Academy of Sciences U.S.A. 25 (1939), 220-224; rev. Journal of Symbolic Logic 5 (1940), 116-117.

(10) — *The consistency of the continuum hypothesis,* Annals of mathematics studies no. 3, 66 pages, Princeton University Press, 1940.

(11) — "Russell's mathematical logic", *The philosophy of Bertrand Russell*, ed. by P.A. Schilpp, pp. 123-153, Evanston and Chicago, 1944.

(12) — "What is Cantor's continuum problem?", The American Mathematical Monthly 54 (1947), 515-525.

(13) — "Über eine bisher noch nicht benützte Erweiterung des finiten

Standpunktes", Dialectica 12 (1958), 280-287. An English translation together with additional notes will appear in Gödel's collected works.

(14) — "Remarks before the Princeton Bicentennial Conference on Problems of Mathematics", *The undecidable*, ed. by Martin Davis (1965) pp. 84-88, New York: Raven Press.

(Gol) GOLDFARB, W.D.: "Logic in the twenties: The nature of the quantifier", Journal of Symbolic Logic 44 (1979), 351-368.

(Gol 1) — "The unsolvability of the Gödel class with identity", Journal of Symbolic Logic 49 (1984), 1237-1252.

(Ha) HASENJAGER, G.: "Eine Bemerkung zu Henkins Beweis für die Vollständigkeit des Prädikatenkalküls", Journal of Symbolic Logic 18 (1953), 42-48.

(H) HOWARD, W.A. "Functional interpretation of bar induction by bar recursion", Compositio Mathematica (1968), 107-124.

(H 1) — "Assignment of ordinals to terms of primitive recursion on functionals of finite type", pp. 443-458 in *Intuitionism and Proof Theory*, ed. by A. Kino, J. Myhill, R.E. Vesley, North-Holland, 1970.

(H 2) — "Ordinal analysis of terms of finite type", Journal of Symbolic Logic 45 (1980) 493-504.

(H 3) — "Ordinal analysis of simple cases of bar recursion", Journal of Symbolic Logic 46 (1981), 17-30.

(HK) HOWARD, W.A. AND KREISEL, G.: "Transfinite induction and bar induction of types 0 and 1, and the role of continuity in intuitionistic analysis", Journal of Symbolic Logic 31 (1966), 325-358.

(J) JONGH, DE, D.H.J.: "A class of intuitionistic connectives", pp. 103-111 in *The Kleene Symposium*,ed. by J. Barwise, North-Holland, 1980.

(Kl) KLEENE, S.C.: "Hierarchies of number-theoretic predicates", Bull. of the Amer. Math. Society 61 (1955), 193-213.

(KV) KLEENE, S.C. AND VESLEY, R.E.: *The foundations of intuitionistic mathematics, especially in relation to recursive functions*, North-Holland, 1965; rev. Journal of Symbolic Logic 31 (1966), 258-261.

(K) KREISEL, G.: "Note on arithmetic models for consistent formulae of the predicate calculus", Fundamenta Mathematicae 37 (1950), 265-285.

(K 1) — "On the interpretation of non-finitist proofs", Journal of Symbolic Logic 16 (1951), 241-267 and 17 (1952) 43-58.

(K 2) — "On the concepts of completeness and interpretation of formal systems", Fundamenta Mathematicae 39 (1952), 103-127.

(K 3) — "Some uses of metamathematics", Brit. Journal of the Philosophy of Science 7 (1956), 161-173.

(K 4) — "Relative consistency proofs (abstract)", Journal of Symbolic Logic 23 (1958), 109-110.

(K 5) — "Functions, ordinals, species", pp. 145-159 in *Logic, Methodology and Philosophy of Science III*, ed. by B. van Rootselaar and J.F. Staal, North-Holland 1968.

(K 6) — "Hilbert's Programme and the search for automatic proof procedures", Springer Lecture Notes in Mathematics 125 (1970), 128-146; rev. Zentralblatt für Mathematik 206 (1971), 277.

(K 7) — "Some reasons for generalizing recursion theory", pp. 139-198 in *Logic Colloquium '69*, ed. by R.O. Gandy, C.M.E. Yates, North-Holland, 1971; rev. Zentralblatt für Mathematik 219 (1972), 17.

(K 8) — "The collected works of Gerhard Gentzen", Journal of Philosophy 68 (1971), 238-265.

(K 9) — "Which number-theoretic problems can be solved in recursive progressions on Π^1_1-paths through O?", Journal of Symbolic Logic 37 (1972), 311-334.

(K 10) — "A notion of mechanistic theory", Synthese 29 (1974), 11-26; rev. Zentralblatt für Mathematik 307 (1976), 02028.

(K 11) — "What have we learnt from Hilbert's second problem?", Proceedings Symp. Pure Math. 28 (1976), 93-130; rev. Zentralblatt für Mathematik 366 (1978) 02018.

(K 12) — "Der unheilvolle Einbruch der Logik in die Mathematik", Acta Philosophica Fennica 28 (1976), 166-187.

(K 13) — "Wie die Beweistheorie zu ihren Ordinalzahlen kam und kommt", Jahresberichte der Deutschen Mathematiker Vereinigung 78 (1977), 177-223.

(K 14) — "Frege's foundations and intuitionistic logic", The Monist 67 (1984), 71-91.

(K 15) — "Einige Erläuterungen zu Wittgensteins Kummer mit Hilbert und Gödel", pp. 295-303 in *Akten des 7. internationalen Wittgenstein Symposiums*, ed. by P. Weingartner and J. Czermak, Hölder-Pichler-Tempsky, Vienna, 1983.

(K 16) — "Proof theory and the synthesis of programs: potential and limitations", Springer Lecture Notes in Computer-Science 203 (1985), 136-150.

(K 17) — "Philosophie: eine Ergänzung der Wissenschaft?", pp. 51-56 in *Akten des 9. internationalen Wittgenstein Symposiums*, ed. by W. Leinfellner und F. Wuketits, Hölder-Pichler-Tempsky, Vienna, 1986.

(KK) Kreisel, G. and Krivine, J.L: *Elements of mathematical logic,* Second Edition, North-Holland, 1971.

(KM) Kreisel, G. and Macintyre, A.: "Constructive logic versus algebraization", pp. 217-260 in *The L.E.J. Brouwer centenary symposium*, ed. by A.J. Troelstra, D. van Dalen, North-Holland, 1982; rev. Zentralblatt für Mathematik 522 (1982), 03046.

(KMS) Kreisel, G., Simpson, S.G. and Mints, G.E.: "The use of abstract language in elementary metamathematics: some pedagogic examples", Springer Lecture Notes in Mathematics 453 (1975) 38-131.

(KT) KREISEL, G. AND TAIT, W.W.: "Finite definability of number-theoretic functions and parametric completeness of equational calculi", Zeitschrift für mathematische Logik und Grundlagen der Mathematik 7 (1961), 28-38.

(KT 1) KREISEL, G. AND TAKEUTI, G.: "Formally self-referential propositions for cut free classical analysis and related systems", Dissertationes mathematicae 118 (1974), 150; rev. Zentralblatt für Mathematik 336 (1977), 02027.

(KT 2) KREISEL, G. AND TROELSTRA, A.S.: "Formal systems for some branches of intuitionistic analysis", Annals of Mathematical Logic 1 (1970), 229-387; an addendum, ibid. 3 (1971), 437-439.

(MS) MANEWITZ, L. AND STAVI, J.: Δ_2^0 operators and alternating sentences in arithmetic", Journal of Symbolic Logic 45 (1980), 144-154.

(M) MOSTOWSKI, A.: *Sentences undecidable in formalized arithmetic,* North-Holland, 1952.

(NH) NERODE, A. AND HARRINGTON, L.: "The work of Harvey Friedman", Notices of the Amer. Math. Society 31 (1984), 563-566; rev. Zentralblatt für Mathematik 588 (1986), 03001.

(PR) POUR EL, M.B. AND RICHARDS, I.: "The wave equation with computable initial data such that its unique solution is not computable", Advances in Mathematics 39 (1981), 215-:239; rev. Journal of Symbolic Logic 47 (1982), 900-902.

(PR 1) — "Non-computability in analysis and physics: a complete determination of the class of non-computable linear operators", Advances in Mathematics 48 (1983), 44-74.

(Re) RENARDEL, G.R.: Theories with type-free application and extended bar induction, Dissertation, Amsterdam 1984.

(Ro) ROBINSON, J.: "Recursive functions of one variable", Proceedings of the Amer. Math. Society 19 (1968), 815-820.

(RS) KREISEL, G.: "Kurt Gödel", Biographical memoirs of Fellows of the Royal Society 26 (1980), 149-224; ibid. 27 (1981), 697 and 28 (1982), 719.

(Sa) SAATY, T.L. (ed.): Lectures on modern mathematics III, Wiley, 1965.

(SH) SCHROEDER-HEISTER, P.: "A natural extension of natural deduction", Journal of Symbolic Logic 49 (1984), 1284-1300.

(S) SPECTOR, C.: "Provably recursive functionals of analysis: a consistency proof of analysis by an extension of principles formulated in current intuitionistic mathematics", Proceedings Symp. Pure Math. 5 (1962), 1-27.

(Sm) SMORYNSKI, C.: "ω-consistency and reflection", pp. 167-181 in Colloque international de logique, CNRS, 1977.

(SV) STEGMÜLLER, W. AND VARGA, M.: *Strukturtypen der Logik,* Springer 1984; rev. Grazer Philosophische Studien 24 (1985), 185-195.

(St) STERN, J. (ed.): *Proceedings of the Herbrand Symposium: Logic Colloquium '81*, North-Holland, 1982.

(Su) SUNDHOLM, G.: "Constructions, proofs and the meaning of the logical constants", Journal of Philosophical Logic 12 (1983), 151-172; rev. Zentralblatt für Mathematik 539 (1985), 03038.

(Ta) TAIT, W.W.: "Intensional interpretation of functionals of finite type", Journal of Symbolic Logic 32 (1967), 198-212.

(TMR) TARSKI, A., MOSTOWSKI, A. AND ROBINSON, R.M.: *Undecidable Theories,* North Holland, 1953.

(T) TROELSTRA, A.S.: "Metamathematical investigations of intuitionistic arithmetic and analysis", Springer Lecture Notes in Mathematics 344, 1973.

(T 1) — "Extended bar induction of type 0", pp. 277-316 in *The Kleene-Symposium*, North-Holland, 1980.

(Tu) TURING, A.M.: "On computable numbers with an application to the Entscheidungsproblem", Proceedings of the London Math. Society 42 (1936), 230-265.

(V) VISSER, A.: Peano's smart children: a provability logical study of systems with built-in consistency. Logic Group Preprint Series, No. 14. Utrecht, 1986.

(We) WEINSTEIN, S.: "The intended interpretation of intuitionistic logic", Journal of Philosophical Logic 12 (1983), 261-270; rev. Zentralblatt für Mathematik 539 (1985), 03037.

(Wi) Scritti su Wittgenstein, Bibliopolis (to appear).

(W) WOJTYLAK, P.: "Collapse of a class of infinite disjunctions in intuitionistic propositional logic", Reports on Mathematical Logic 16 (1982), 37-49.

(W 1) — "A recursive theory for the $\{\neg, \wedge, \vee, \rightarrow, \circ\}$ fragment of intuitionistic logic", Ibid. 18 (1984), 3-35.

(Z) ZERMELO, E.: "Über Stufen der Quantifikation und die Logik des Unendlichen", Jahresberichte der Deutschen Mathematikervereinigung 41 (1932), 85-88.

(Z 1) — "Grundlagen einer allgemeinen Theorie der mathematischen Satzsysteme", Fundamenta Mathematicae 25 (1935), 136-146.

LIST OF ACTIVE PARTICIPANTS

Curt Christian, Institut für Logistik, University of Vienna

Martin D. Davis, Department of Mathematics, New York University

John W. Dawson, The Institute for Advanced Studies, School of Mathematics, Princeton

Solomon Feferman, Department of Mathematics, Stanford University

Stephen C. Kleene, Department of Mathematics, University of Wisconsin, Madison

Georg Kreisel, Stanford University and Institut für Wissenschaftstheorie, International Research Center, Salzburg

Sir Karl R. Popper, London School of Economics and L. Boltzmann Institut, Vienna

Leopold Schmetterer, Austrian Academy of Sciences, Vienna

Dana Scott, Department of Computer Science, Carnegie-Mellon University, Pittsburgh

Olga Taussky-Todd, A.P. Sloan Laboratory of Mathematics and Physics, California Institute of Technology

Paul Weingartner, Institut für Wissenschaftstheorie, International Research Center Salzburg and Institut für Philosophie, Universität Salzburg

Stampato nel maggio 1987
nelle officine grafiche
della Grafitalia s.r.l.
Cercola (Napoli)